Hematology: 101 Morphology Updates

Hematology

101 Morphology Updates

Barbara J. Bain MBBS, FRACP, FRCPath
Professor of Diagnostic Haematology
St Mary's Hospital Campus of Imperial College London

and

Honorary Consultant Haematologist
St Mary's Hospital
Imperial College Healthcare NHS Trust
London

Registered Offices
John Wiley & Sons, Inc., 111 River Street, Hoboken, NJ 07030, USA
John Wiley & Sons Ltd, The Atrium, Southern Gate, Chichester, West Sussex, PO19 8SQ, UK

For details of our global editorial offices, customer services, and more information about Wiley products visit us at www.wiley.com.

Wiley also publishes its books in a variety of electronic formats and by print-on-demand. Some content that appears in standard print versions of this book may not be available in other formats.

Library of Congress Cataloging-in-Publication Data Applied for
Hardback: 9781394179817

Cover Design: Wiley
Cover Image: © Roger Sutcliffe/Getty Images

Set in 9.5/12.5pt STIXTwoText by Straive, Pondicherry, India

Contents

Preface

This book is based on an ongoing series of Morphology Updates that have been published monthly in the *American Journal of Hematology* since 2008. When necessary, cases have been updated and a test yourself section has been added. The aim of the book is to bring the importance of hematological morphology to a wider readership. In selecting cases for inclusion, preference has been given to those where microscopy was crucial for diagnosis and where there is a generalizable message.

Note to the reader

Unless otherwise stated, all photomicrographs have been stained with a May–Grünwald–Giemsa or similar Romanowsky-type stain. Photography has generally been with a ×100 objective but sometimes ×50 or other magnification.

Acknowledgements

I wish to acknowledge the Editor of the *American Journal of Hematology*, Carlo Brugnara, whose idea it was to gather Morphology Updates into book form and thus disseminate them more widely.

The major role of the co-authors of the original reports is gratefully acknowledged. They are listed alphabetically below, in groups according to their affiliation at the time of writing the initial update, and the original article is cited with each Update.

Co-authors who are or were, at the time of writing the Update, attached to Imperial College London or Imperial College Healthcare NHS Trust (St Mary's Hospital, Hammersmith Hospital and Charing Cross Hospital)

Saad Abdalla, Syed Ahmed, Jane Apperley, Marc Arca, Maria Atta, Anna Austin, Peter Bain, Vandana Bharadwaj, Eimear Brannigan, Loretta Brown, Victoria Campbell, Aristeidis Chaidos, Subarna Chakravorty, Lynda Chapple, Kan Cheung, Lucy Cook, Nichola Cooper, Christina Crossette-Thambiah, Josu De La Fuente, Simona Deplano, Nadine Farah, Rashpal Flora, Rodney Foale, Emma Fosbury, Jacob Grinfeld, Kamala Gurung, Sophie Hanina, Amanda Hann, Andrew Hastings, Marc Heller, Leena Karnik, Mark Layton, Thomas Lofaro, Kirstin Lund, Asad Luqmani, Sasha Marks, Philippa May, Dragana Milojkovic, Audrey Morris, Jane Myburgh, Elizabet Nadal-Melsió, David Nam, Akwasi Osei-Yeboah, Bella Patel, Jiří Pavlů, Lorry Phelan, Amin Rahemtulla, Edward Renaudon-Smith, John Riches, Lynn Robertson, Megan Rowley, Gayathriy Sivaguru, Michael Spencer-Chapman, Sree Sreedhara, Matthew Stubbs, Sarmad Toma, James Uprichard, Lewis Vanhinsbergh, Vanlata Varu and Eva Yebra-Fernandez.

Co-authors attached to other UK hospitals and institutions

Bahaa Al-Bubseree, Beatson West of Scotland Cancer Centre, Glasgow;
Magda Al Obaidi, West Middlesex University Hospital, Isleworth, London;
Hannah Al-Yousuf, North Middlesex University Hospital, London;
Philip Ancliff, Great Ormond Street Hospital for Children, London;
Anna Babb, West Middlesex University Hospital, Isleworth, London;
Linda Barton, University Hospitals of Leicester NHS Trust, Leicester;
Tanya Bernard, Ashford and St Peter's Hospitals NHS Foundation Trust, Chertsey;
Manju Bhavnani, Royal Albert Edward Infirmary, Wigan;
Kieran Burton, Wycombe Hospital, Buckinghamshire Healthcare NHS Trust, High Wycombe;
Carolyn Campbell, Oxford Genetics Laboratories, Oxford University Hospitals NHS Trust, Oxford;
Wei Yee Chan, University College London Hospitals NHS Foundation Trust, London;
Peter Chiodini, Hospital for Tropical Diseases, London;

Barnaby Clark, King's College and King's College Hospital, London;
Nicholas Cross, Wessex Regional Genetics Laboratory, Salisbury District Hospital, Salisbury;
Helen Eagleton, Wycombe Hospital, Buckinghamshire Healthcare NHS Trust, High Wycombe;
Emilia Escuredo, St Thomas' Hospital, London;
Rachel Farnell, Wycombe Hospital, Buckinghamshire Healthcare NHS Trust, High Wycombe;
Nicholas Fordham, St Helier Hospital, Carshalton, London;
Niharendu Ghara, North Middlesex University Hospital, London;
Kirsteen Harper, Beatson West of Scotland Cancer Centre, Glasgow;
David Hopkins, Beatson West of Scotland Cancer Centre, Glasgow;
Ann Hunter, University Hospitals of Leicester NHS Trust, Leicester;
Vishal Jayakar, Kingston Hospital, Kingston upon Thames, London;
Rosie Jones, Borders General Hospital, Melrose;
Andrew Keenan, North Devon District Hospital, Barnstaple;
Ahmad Khoder, West Middlesex University Hospital, Isleworth, London;
May-Jean King, NHS Blood and Transplant, Bristol;
Victoria Kronsten, King's College Hospital, London;
Alison Laing, Beatson West of Scotland Cancer Centre, Glasgow;
Mike Leach, Beatson West of Scotland Cancer Centre, Glasgow;
Christine Liu, West Middlesex University Hospital, Isleworth, London;
John Luckit, North Middlesex University Hospital, London;
Caitlin MacDonald, North Devon District Hospital, Barnstaple;
Louisa McIlwaine, Beatson West of Scotland Cancer Centre, Glasgow;
Francis Matthey, Chelsea and Westminster Hospital, London;
Sajir Mohamedbhai, North Middlesex University NHS Trust, London;
Veselka Nikolova, Royal Marsden Hospital, Sutton;
Simon O'Connor, Royal Marsden Hospital, Sutton;
Katrina Parsons, Beatson West of Scotland Cancer Centre, Glasgow;
Sophie Portsmore, North Middlesex University Hospital, London;
Clare Rees, Frimley Park Hospital, Camberley;
Debbie Shawcross, King's College Hospital, London;
Giulia Simini, West Middlesex University Hospital, Isleworth, London;
Wenchee Siow, Wycombe Hospital, Buckinghamshire Healthcare NHS Trust, High Wycombe;
Dean Smyth, Beatson West of Scotland Cancer Centre, Glasgow;
Simon Stern, St Helier Hospital, Carshalton, London;
John Swansbury, The Royal Marsden Hospital, Sutton;
Sabita Uthaya, Chelsea and Westminster Hospital, London;
Godhev Vijay, King's College Hospital, London;
Barbara Wild, King's College Hospital, London;
Ke Xu, University College London Hospitals NHS Foundation Trust, London;
and Anne Yardumian, North Middlesex University Hospital, London.

Co-authors from Australia
Alan Mills, Bendigo, Victoria (now deceased);
Robyn Wells, Princess Alexandra Hospital, Woolloongabba, Queensland;
Bronwyn Williams, Royal Brisbane and Women's Hospital, Herston, Queensland;
and Hui Sien Tay, Bendigo, Victoria.

Co-author from France
Jean-Baptiste Rieu, Cancer University Institute of Toulouse Oncopole, Toulouse.

Co-author from India
Biswadip Hazarika, Batra Hospital and Medical Research Centre, New Delhi.

Co-authors from Iraq
Abbas Hashim Abdulsalam, Al-Yarmouk Teaching Hospital, Baghdad;
Abdulsalam Hatim, National Center for Hematology, Baghdad;
Zead Ibrahim, Al-Yarmouk Teaching Hospital, Baghdad;
Mohammed Khamis, Al-Khadimiya Teaching Hospital, Baghdad;
and Nafila Sabeeh, Al-Yarmouk Teaching Hospital, Baghdad.

Co-authors from Italy
Francesca Barducchi, Anatomical Pathology, ASL2 Liguria;
Enrico Cappelli, Clinical Pathology, ASL2 Liguria;
Daniele Delli Carri, Azienda Ospedaliera G. Rummo, Benevento;
Maurizio Fumi, Azienda Ospedaliera G. Rummo, Benevento;
Brisejda Koroveshi, Clinical Pathology, ASL2 Liguria;
Lorella Lanza, Anatomical Pathology, ASL2 Liguria;
Flavia Lillo, Clinical Pathology, ASL2 Liguria;
Ylenia Pancione, Azienda Ospedaliera G. Rummo, Benevento;
Vincenzo Rocco, Azienda Ospedaliera G. Rummo, Benevento;
Silvia Sale, Azienda Ospedaliera G. Rummo, Benevento;
and Ezio Venturino, Anatomical Pathology, ASL2 Liguria.

Co-author from Kuwait
Hassan A. Al-Jafar, Amiri Hospital, Kuwait City.

Co-authors from Portugal
Rui Barreira, Instituto Português de Oncologia de Lisboa Francisco Gentil (IPOLFG), Lisbon;
José Cortez, Instituto Português de Oncologia de Lisboa Francisco Gentil (IPOLFG), Lisbon;
Filipa Fernandes, Instituto Português de Oncologia de Lisboa Francisco Gentil (IPOLFG), Lisbon;
Rita Ramalho, Instituto Português de Oncologia de Lisboa Francisco Gentil (IPOLFG), Lisbon;
and Margarida Silveira, Instituto Português de Oncologia de Lisboa Francisco Gentil (IPOLFG), Lisbon.

Co-authors from Spain
Beatriz Bua, Hospital Clínico Universitario de Salamanca, Salamanca;
Félix Cadenas, Hospital Clínico Universitario de Salamanca, Salamanca;
and María Campelo, Hospital Clínico Universitario de Salamanca, Salamanca.

Abbreviations

ADAMTS13	a disintegrin and metalloprotease domain with thrombospondin type 1 motif, member 13
AIDS	acquired immune deficiency syndrome
ALL	acute lymphoblastic leukemia
AML	acute myeloid leukemia
ANAE	α naphthyl acetate esterase
APL	acute promyelocytic leukemia
APTT	activated partial thromboplastin time
ATLL	adult T-cell leukemia/lymphoma
ATP	adenosine triphosphate
ATRA	all-*trans*-retinoic acid
BIA-ALCL	breast implant-associated anaplastic large cell lymphoma
BPDCN	blastic plasmacytoid dendritic cell neoplasm
B-PLL	B-cell prolymphocytic leukemia
CD	cluster of differentiation
CLL	chronic lymphocytic leukemia
CMV	cytomegalovirus
CNL	chronic neutrophilic leukemia
COVID-19	corona virus disease 2019
CVAD	cyclophosphamide, vincristine, doxorubicin (Adriamycin) and dexamethasone
2,3-DPG	2,3-diphosphoglycerate
DIC	disseminated intravascular coagulation
DNA	deoxyribonucleic acid
EBER	Epstein–Barr virus-encoded small RNA
EBV	Epstein–Barr virus
EDTA	ethylenediaminetetra-acetic acid
EMA	eosin-5′-maleimide
FAB	French–American–British (classifications of hematological neoplasms)
FEU	fibrinogen equivalent units
FISH	fluorescence *in situ* hybridization
G6PD	glucose-6-phosphate dehydrogenase
G-CSF	granulocyte colony-stimulating factor
GVHD	graft-versus-host disease
H&E	hematoxylin and eosin

Hb	hemoglobin concentration
Hct	hematocrit
HELLP	hemolysis, elevated liver enzymes and low platelet count (syndrome)
HIV	human immunodeficiency virus
HLA	histocompatibility locus antigen
HPLC	high performance liquid chromatography
HTLV-1	human T-cell lymphotropic virus 1
Ig	immunoglobulin
IL	interleukin
IRF4	interferon regulatory factor 4
ITD	internal tandem duplication
ITP	immune thrombocytopenic purpura/autoimmune thrombocytopenic purpura
LDH	lactate dehydrogenase
LE	lupus erythematosus
MBCL	monoclonal B-cell lymphocytosis
MCF	mean cell fluorescence
MCH	mean cell hemoglobin
MCHC	mean cell hemoglobin concentration
MCV	mean cell volume
MDS	myelodysplastic syndrome/s
mRNA	messenger ribonucleic acid
MUM1	multiple myeloma oncogene 1
NADPH	nicotinamide adenine dinucleotide phosphate
NK	natural killer (cell)
NRBC	nucleated red blood cell/s
P5'N	pyrimidine 5' nucleotidase
PAS	periodic acid–Schiff (reaction)
PCH	paroxysmal cold hemoglobinuria
PCR	polymerase chain reaction
PLT	platelet
PT	prothrombin time
RBC	red blood cell count
RDW	red cell distribution width
RT-PCR	reverse transcriptase polymerase chain reaction
SARS-CoV-2	severe acute respiratory distress syndrome corona virus 2
SLE	systemic lupus erythematosus
Sp	spectrin
SUV	standardized uptake value
TA-TMA	transplant-associated thrombotic microangiopathy
TNCC	total nucleated cell count
T-PLL	T-cell prolymphocytic leukemia
TTP	thrombotic thrombocytopenic purpura
VEGF	vascular endothelial growth factor
WBC	white blood cell count
WHO	World Health Organization

1   Malaria – one swallow makes a summer

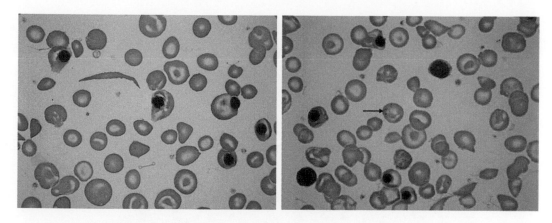

A Nigerian woman who was 37 weeks' pregnant, recently arrived in the UK, presented to a general practitioner with tiredness and dyspnea. Her automated full blood count showed a 'white cell count' of 112×10^9/l, Hb 55 g/l, MCV 101 fl and platelet count 471×10^9/l. Examination of a blood film showed that the elevated 'white cell count' was due to large numbers of nucleated red blood cells and the true white cell count was 10.7×10^9/l. In addition, the film showed features of sickle cell disease (left image) and high performance liquid chromatography showed hemoglobin S as the major hemoglobin with no hemoglobin A being present. The lack of microcytosis indicated that the diagnosis was sickle cell anemia (SS) rather than compound heterozygosity for hemoglobin S and β^0 thalassemia. The patient, when questioned, stated that she had sickle cell trait and denied any knowledge of a diagnosis of sickle cell anemia.

Since the Hb was somewhat lower than expected (although compatible with sickle cell anemia in late pregnancy) the blood film was further examined to try to identify any other factors contributing to the anemia. There was marked polychromasia. There were no hypersegmented neutrophils and the MCV was considered compatible with the reticulocytosis. Unexpectedly, a *Plasmodium falciparum* ring form was detected (right). A careful search of the film disclosed a total of four parasites. A Giemsa stain showed Maurer clefts and immunological tests for an antigen specific to *P. falciparum* confirmed the diagnosis. Further questioning of the patient, who was afebrile, disclosed that 3 weeks earlier she had suspected that she had malaria and had taken a single tablet of Fansidar (pyrimethamine plus sulfadoxine) plus a paracetamol tablet. Further appropriate treatment for falciparum malaria was given.

It is stated that 'one swallow does not a summer make' but the detection of a single parasite does permit a diagnosis of malaria.

Original publication: Bain BJ (2012) Malaria – one swallow makes a summer. *Am J Hematol*, **87**, 190.

Hematology: 101 Morphology Updates, First Edition. Barbara J. Bain.
© 2023 John Wiley & Sons Ltd. Published 2023 by John Wiley & Sons Ltd.

2 The significance of irregularly contracted cells and hemighosts in sickle cell disease

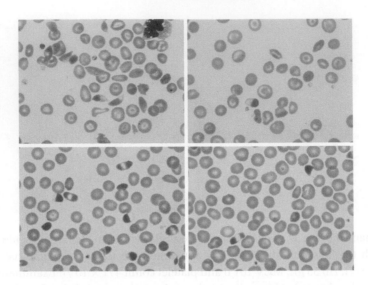

The presence of considerable numbers of irregularly contracted cells and hemighosts in sickle cell disease may serve as a warning of severe sickle crisis with significant hypoxia. These two cases demonstrate this association.

The first patient was a 52-year-old woman with sickle cell/β^0 thalassemia who presented with generalized bone pain. She was given intravenous fluids, antibiotics and analgesics but, despite treatment, developed respiratory failure with a PO_2 of 7.0 kPa. Her blood count showed WBC 10.0×10^9/l, Hb 72 g/l, MCV 63.9 fl and platelets 257×10^9/l. In addition to the usual features of sickle cell disease, her blood film showed hemighost cells in which the hemoglobin was retracted to one side of the erythrocyte (top right image). This phenomenon was also observed in cells showing evidence of hemoglobin polymerization – cells with pointed ends and sometimes a gentle curve (top left). She declined continuous positive airway pressure and was commenced on high flow oxygen by nasal cannula, to which she responded well.

The second patient was a young man with sickle cell anemia. He suffered a cardiac arrest and became severely hypoxic. His blood count then showed WBC 26.5×10^9/l, Hb 101 g/l, MCV 88.7 fl and platelet count 110×10^9/. His blood film showed numerous hemighosts with hemoglobin retracted to one side of the cell or to both ends of an elongated cell (bottom images). In addition, there were irregularly contracted cells, some of which were somewhat angular, suggesting that polymerization was occurring (bottom left).

Irregularly contracted cells and hemighosts can be a warning sign of severe hypoxia and worsening crisis in patients with sickle cell disease. This blood film observation reflects the increased percentage of dense cells demonstrable on density gradient analysis in patients with sickle cell crisis.[1]

Original publication: Siow W, Matthey F and Bain BJ (2017) The significance of irregularly contracted cells and hemighosts in sickle cell disease. *Am J Hematol*, **92**, 966–967.

Reference

1 Fabry ME and Kaul DK (1991) Sickle cell vaso-occlusion. *Hematol Oncol Clin North Am*, **5**, 375–398.

Hematology: 101 Morphology Updates, First Edition. Barbara J. Bain.
© 2023 John Wiley & Sons Ltd. Published 2023 by John Wiley & Sons Ltd.

3 Striking dyserythropoiesis in sickle cell anemia following an aplastic crisis

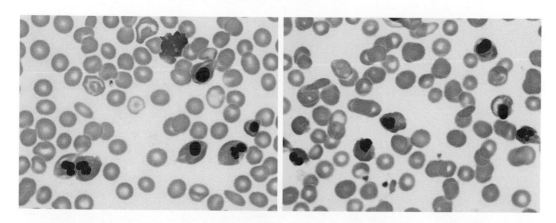

A 7-year-old boy with clinically mild sickle cell anemia was transferred to our pediatric intensive care unit from his local hospital. He had presented with fatigue, had developed respiratory distress and had been found to have metabolic acidosis, markedly elevated troponin and severe anemia. His Hb was 20 g/l, with an inadequate reticulocyte response (reticulocyte count 32×10^9/l). He had been transfused 20 ml/kg of O RhD-negative packed red cells and had been commenced on broad-spectrum antimicrobials prior to transfer. On arrival he was hemodynamically stable and self-ventilating.

A chest X-ray demonstrated right-sided consolidation. An echocardiogram showed mild mitral regurgitation and tricuspid regurgitation with a dilated left ventricle and mildly reduced ventricular function. Infection screening including urine legionella and pneumococcal antigen, blood cultures, urine cultures and viral respiratory screen were all negative. The patient was intubated and ventilated shortly after arrival in the pediatric intensive care unit. He improved clinically over the next few days, following further red cell transfusion and escalation of antibiotics, and was extubated. Serum troponin gradually normalized. As he recovered, there was an outpouring of nucleated red blood cells into the peripheral blood, with striking dyserythropoiesis. Erythroblasts showed nuclear lobulation, basophilic stippling, detached nuclear fragments, occasional binucleated forms and occasional mitotic figures (images).

Parvovirus B19 serology, IgG and IgM, was positive, confirming recent parvovirus infection. Thus the suspected diagnosis of parvovirus-induced severe aplastic crisis was confirmed, this being complicated by pneumonia and by cardiac ischemia secondary to severe anemia.

Dyserythropoiesis has many causes, including hemolytic anemia. When erythropoiesis is very active – 'stress erythropoiesis' – dyserythropoiesis can be striking, as in this patient. It is important to be aware of the many potential causes of dyserythropoiesis in order to avoid misdiagnosis.

Original publication: Austin A, Lund K and Bain BJ (2019) Striking dyserythropoiesis in sickle cell anemia following an aplastic crisis. *Am J Hematol*, **94**, 378.

Hematology: 101 Morphology Updates, First Edition. Barbara J. Bain.
© 2023 John Wiley & Sons Ltd. Published 2023 by John Wiley & Sons Ltd.

4 A normal mean cell volume does not exclude a diagnosis of megaloblastic anemia

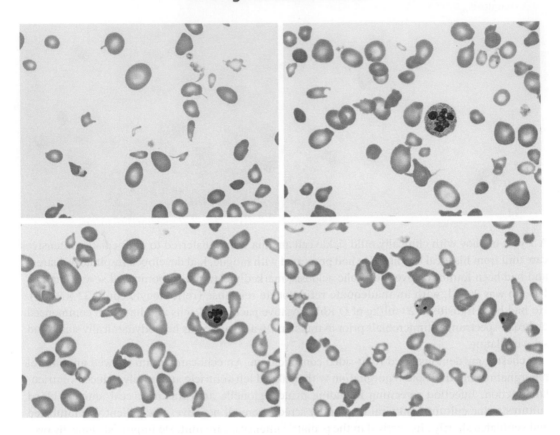

A 61-year-old male was referred to the emergency department by his general practitioner because of pancytopenia. This had been noted several months previously but the patient had refused detailed investigation or treatment. He was noted to have hyperpigmentation of his hands with palmer and plantar freckling. His blood count showed WBC 1.9×10^9/l, neutrophils 0.8×10^9/l, RBC 1.69×10^{12}/l, Hb 42 g/l, MCV 81.5 fl, MCH 28 pg, MCHC 343 g/l, red cell distribution width (RDW) 35.5% (normal range 10–16), platelets 47×10^9/l and reticulocytes 44.5×10^9/l. His blood film (all images) showed marked anisocytosis and poikilocytosis, red cell fragments, macrocytes, oval macrocytes, teardrop poikilocytes and Howell–Jolly bodies (bottom right). Biochemical tests showed lactate dehydrogenase 5594 iu/l, creatinine 84 μmol/l, bilirubin 59 μmol/l and ferritin 257 μg/l. Serum haptoglobin was 0 g/l (0.52–2.24). A coagulation screen was normal.

The marked red cell fragmentation raised the suspicion of thrombotic thrombocytopenic purpura (TTP) but a careful inspection of the blood film suggested an alternative diagnosis. Hypersegmented neutrophils were present (top right) and there were circulating megaloblasts, some of which showed dyserythropoietic features (bottom left). In addition, oval macrocytes and teardrop poikilocytes are very typical of megaloblastic anemia. Serum vitamin B_{12} was found to be <148 ng/l (160–800) and serum folate was normal at 7.1 μg/l. Parietal cell, but not intrinsic factor, antibodies were identified. ADAMTS13 was 26.3%. Results were retrieved from the hospital that the patient had attended several months earlier. These had shown Hb 79 g/l, MCV 106 fl and vitamin B_{12} 100 ng/l (191–663) with a normal serum folate.

A diagnosis of pernicious anemia was made and the patient then consented to treatment with parental hydroxocobalamin. Within 2 weeks his results were WBC 9.8×10^9/l, Hb 87 g/l, MCV 81.2 fl, neutrophils 6.4×10^9/l, platelets 603×10^9/l and reticulocytes 129×10^9/l.

This patient shows the paradoxical fall of the MCV that can occur as the Hb falls in untreated megaloblastic anemia. It should be noted that although the MCV has fallen from an elevated level to within the normal range the RDW is very high. This is the result of the increased red cell fragmentation that occurs as megaloblastosis worsens and is reflected in the red cell size plot from an automated counter (Alinity h-series) (image below). It will be seen that although there are considerable numbers of macrocytes and the mean volume is normal, the mode of cell size is 50 fl with many even smaller red cell fragments.

Schistocytes and keratocytes are not infrequent in severe megaloblastic anemia and can lead to a suspicion of microangiopathic hemolytic anemia.[1] It is important that such patients are not misdiagnosed as having TTP and that the possibility of megaloblastic anemia is not discounted because the MCV is normal.

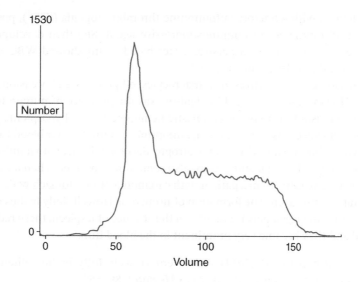

Original publication: Gurung K and Bain BJ (2021) A normal mean cell volume does not exclude a diagnosis of megaloblastic anemia. *Am J Hematol*, **96**, 1706–1707.

Reference

1 Bain BJ (2010) Schistocytes in megaloblastic anemia. *Am J Hematol*, **85**, 599.

5 Prominent Howell–Jolly bodies when megaloblastic anemia develops in a hyposplenic patient

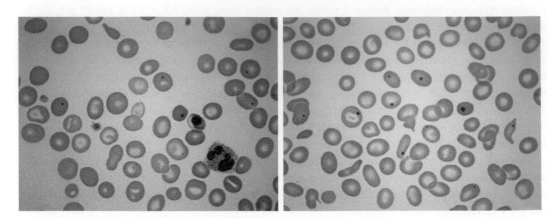

A middle-aged female with refractory autoimmune thrombocytopenia ('ITP'), post-splenectomy, was prescribed azathioprine as an immunosuppressive agent. She then developed anemia and macrocytosis. Her thrombocytopenia persisted. Her blood count showed WBC 8.1×10^9/l, Hb 95 g/l, MCV 109 fl and platelet count 23×10^9/l.

Her blood film showed macrocytosis, oval macrocytes, elliptocytes and occasional erythroblasts and myelocytes. However, the most striking feature was the presence of numerous Howell–Jolly bodies (images). This is a feature of megaloblastic hemopoiesis in a hyposplenic patient. In this patient the megaloblastosis was the result of azathioprine therapy. Similar blood films can be seen when patients with celiac disease with splenic atrophy develop folic acid or vitamin B_{12} deficiency. In the past, these morphological features were also sometimes observed when a total gastrectomy and splenectomy were performed in a patient with carcinoma of the stomach with no replacement vitamin B_{12} therapy being given. The formation of numerous Howell–Jolly bodies occurs as a dyserythropoietic feature in megaloblastic anemia. In the absence of a spleen, these red cell inclusions are not removed and they become very prominent in the blood film.

Original publication: Bain BJ (2014) Prominent Howell–Jolly bodies when megaloblastic anemia develops in a hyposplenic patient. *Am J Hematol*, **89**, 852.

6   A ghostly presence – G6PD deficiency

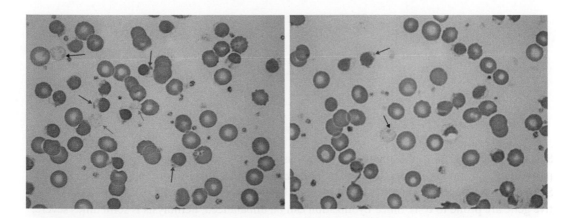

Attention has been drawn to the importance of irregularly contracted cells and hemighosts in suggesting the diagnosis of hemolysis due to glucose-6-phosphate dehydrogenase (G6PD) deficiency.[1] The current patient, a man of African ancestry, suffered an acute fall of his Hb from normal to 65 g/l following exposure to rasburicase. The MCHC was 362 g/l. Rasburicase had been administered prior to chemotherapy for Burkitt lymphoma. Examination of the blood film showed irregularly contracted cells and hemighosts, some containing visible Heinz bodies. In addition, there were considerable numbers of ghost cells – empty red cell membranes bereft of hemoglobin (images). Within some of the ghost cells, Heinz bodies were visible (black arrows, Heinz bodies in ghost cells; blue arrows, other ghost cells; red arrows, Heinz bodies in hemighost cells). The increased MCHC is attributable to the presence of irregularly contracted cells but is not specific; it is also a feature of spherocytic anemias.

G6PD assay was performed and was normal. Subsequent assay, after resolution of the hemolytic episode, confirmed G6PD deficiency, A− variant. An assay is particularly likely to be normal shortly after acute hemolysis in the A− variant because of the normal level of the enzyme in reticulocytes.

The history in this patient is strongly suggestive of G6PD deficiency and confirmatory evidence is provided by the blood film. Together, they avoid the risk of misinterpretation of the initially normal assay. The presence of ghost cells indicates recent acute intravascular hemolysis.

Original publication: Bain BJ (2010) A ghostly presence. *Am J Hematol*, **85**, 271.

Reference

1 Bain B (2008) Sudden onset of jaundice in a Sardinian man. *Am J Hematol*, **83**, 810.

7 G6PD deficiency in patients identified as female

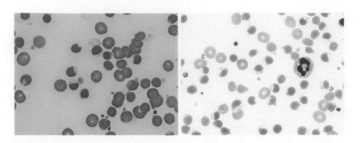

Symptomatic glucose-6-phosphate dehydrogenase (G6PD) deficiency, being X-linked, is seen particularly in males. Cases in female homozygotes are uncommon but well recognized. Symptomatic hemolysis can also occur in female heterozygotes since cells that express the defective gene are prone to lysis. Since Lyonization can be unbalanced, hemolysis is sometimes severe. There are also other uncommon circumstances when G6PD deficiency leads to clinically apparent hemolysis in females or patients identified as female. Recognition of this possibility requires close liaison between clinical and laboratory staff.

The left image is the blood film of a 3-year-old Syrian girl who presented with acute hemolysis 3 days after eating falafel. Her blood count showed an Hb of 74 g/l and an MCHC of 375 g/l. The image shows irregularly contracted cells and numerous 'blister cells' or 'hemighosts'. In addition, there is one cell that is virtually devoid of hemoglobin, a 'ghost cell'. G6PD was 2.2 u/g Hb (normal range 6.3–11.2). The increased MCHC reflects the presence of numerous irregularly contracted cells. Because the hemolysis was unusually severe for a female, DNA analysis was performed. This showed homozygosity for G6PD c. 653C>T; p. Ser218Phe, also known as G6PD Mediterranean. The patient's 5-year-old brother presented simultaneously with an Hb of 70 g/l and a G6PD assay of 2 u/g Hb. The patient's father, who was known to be G6PD deficient, had milder hemolysis. Falafel are traditionally made with chick peas but are sometimes made with fava beans (broad beans) or with a mixture of the two. Patients and physicians may not be aware of this possibility.

The right image is of the blood film of an adult South Asian patient identified in hospital records as 'female', who presented with fever and symptomatic anemia. It shows similar features to the first patient with numerous irregularly contracted cells and blister cells. In addition, Heinz bodies are apparent, precipitated within the otherwise empty area of cytoplasm within blister cells. Hb was 70 g/l. A G6PD assay confirmed deficiency. Again, the hemolysis was unusually severe for a female and further enquiries were made. It was discovered that the patient was a trans female, genetically male but choosing to identify in medical records as female.

Other circumstances in which G6PD deficiency leading to symptomatic hemolysis of unexpected severity in a female that have been reported include Turner syndrome (female patients with a single X chromosome) and females who have been transplanted with bone marrow from a G6PD-deficient male. As it becomes possible in various jurisdictions for trans persons to legally change their gender in official records, diagnostic conundrums are likely to arise. Close liaison between clinical and laboratory staff is essential for these patients and also for those who have had a bone marrow transplant that is unknown to the laboratory as it was performed in another hospital.

Original publication: Bain BJ, Myburgh J, Lund K and Chaidos A (2023) G6PD deficiency in patients identified as female. *Am J Hematol*, **98**, 359–360.

Hematology: 101 Morphology Updates, First Edition. Barbara J. Bain.
© 2023 John Wiley & Sons Ltd. Published 2023 by John Wiley & Sons Ltd.

8 The cause of sudden anemia revealed by the blood film

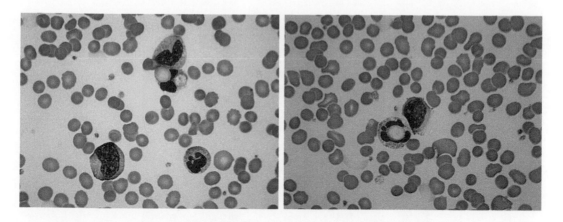

A girl aged 3 years 6 months of Indian ethnic origin presented with a 2-day history of fever and sore throat. She had also been passing red urine for the previous 24 hours. Her blood count showed WBC 27.5×10^9/l, RBC 2.76×10^{12}/l, Hb 81 g/l, Hct 0.22 l/l, MCV 80 fl, MCH 29.6 pg and MCHC 370 g/l. The increased MCHC suggested the presence of spherocytes, irregularly contracted cells or some other hyper-dense cells. A blood film confirmed the presence of numerous spherocytes and in addition was leucoerythroblastic and showed red cell agglutination and erythrophagocytosis by neutrophils (images). Other neutrophils had large vacuoles, similar in size to an erythrocyte, containing amorphous debris. There were also atypical lymphocytes, which appeared reactive. Lactate dehydrogenase and bilirubin were increased. The reticulocyte count was initially normal (46×10^9/l) but subsequently rose. The combination of marked spherocytosis, red cell agglutinates and erythrophagocytosis was considered strongly suggestive of paroxysmal cold hemoglobinuria (PCH) and confirmatory tests were performed. A direct antiglobulin test was positive for complement (+++) and negative for IgG. A Donath–Landsteiner test was positive. Red cell transfusion was required and full recovery had occurred by 3 weeks.

Although the Donath–Landsteiner antibody is an IgG antibody it is usual for the direct antiglobulin test in PCH to show complement only since the antibody that binds to the cell and fixes complement in the cold detaches from the red cell membrane on warming. Confirmation of the diagnosis is by demonstration of the presence of an anti-P antibody and biphasic hemolysis. However, the blood film appearances are highly characteristic and permit a rapid presumptive diagnosis.

Original publication: Bharadwaj V, Chakravorty S and Bain BJ (2011) The cause of sudden anemia revealed by the blood film. *Am J Hematol*, **87**, 520.

Hematology: 101 Morphology Updates, First Edition. Barbara J. Bain.
© 2023 John Wiley & Sons Ltd. Published 2023 by John Wiley & Sons Ltd.

9 Choreo-acanthocytosis

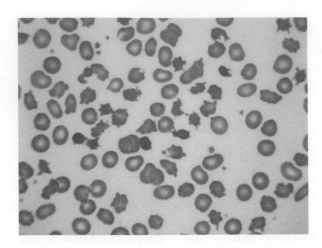

A 38-year-old Pakistani patient presented with a 3-year history of facial grimacing, dysphagia, dysarthria and stutter. More recently she had developed mild chorea. On examination, she had facial tics, eye twitching, excessive blinking, tongue protrusions, dystonic opening of the mouth and choreiform movement of the limbs, particularly on the right side.[1] There was no family history of neurological disease but her parents were first cousins. Her blood count was normal but a blood film showed acanthocytosis (image). A diagnosis of choreo-acanthocytosis was confirmed by demonstration of reduced chorein expression in red cell membranes.

Choreo-acanthocytosis is one of four rare neurological syndromes with movement disorders linked to atrophy of the basal ganglia, acanthocytosis and normal β lipoproteins, designated collectively neuroacanthocytosis (table below). The common link between the neurological and erythroid abnormality is an abnormality of cell membranes. Detection of acanthocytes in a blood film is diagnostically useful.

Syndrome	Mutated gene and inheritance	Clinicopathological features
Choreo-acanthocytosis	*VPS13A*, autosomal recessive	Adult-onset progressive neurodegeneration, myopathy, often epilepsy
McLeod phenotype	*KX*, X-linked recessive	Adult-onset progressive neurodegeneration, myopathy, cardiomyopathy, weak or absent expression of Kell antigens
Huntingdon-like disease 2*	*JPH3*, autosomal dominant	Adult-onset progressive neurodegeneration
Pantothenate-kinase associated neurodegeneration*	*PANK2*, autosomal recessive	Childhood-onset progressive neurodegeneration, pallidal degeneration, sometimes retinitis pigmentosa

* Some cases have acanthocytes

Hematology: 101 Morphology Updates, First Edition. Barbara J. Bain.
© 2023 John Wiley & Sons Ltd. Published 2023 by John Wiley & Sons Ltd.

Original publication: Bain BJ and Bain PG (2013) Choreo-acanthocytosis. *Am J Hematol*, **88**, 712.

Reference

1 Sokolov E, Schneider SA and Bain PG (2012) Chorea-acanthocytosis. *Pract Neurol*, **12**, 40–43.

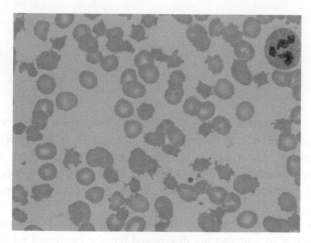

Another image of the patient's blood film showing a neutrophil and acanthocytic red cells.

10 Lead poisoning

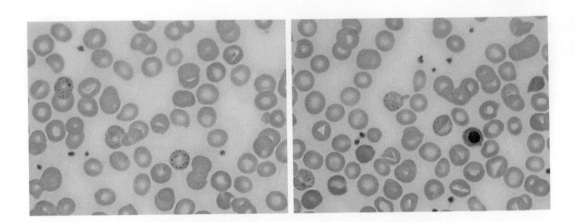

A 50-year-old Indian woman newly arrived in the UK from India had a 9-month history of anorexia, nausea, constipation and dyspnea. Her blood count showed WBC 9×10^9/l, Hb 83 g/l, MCV 85 fl, platelet count 249×10^9/l and reticulocyte count 281×10^9/l. Her blood film showed prominent basophilic stippling (left image), involving even an occasional circulating nucleated red blood cell (right). Mild poikilocytosis and some polychromatic macrocytes were also noted.

Examination of the patient showed a lead line on her gums and a serum lead of 82 μg/100 ml. The source of lead in this patient was not discovered but she recovered on moving to England. Acknowledged sources of lead include Indian cosmetics and ayurvedic medicines, not only those available in India but also those manufactured in the USA and sold via the internet.[1]

Hematological effects of lead poisoning include a hypochromic microcytic anemia with sideroblastic erythropoiesis, a hemolytic anemia and a leucoerythroblastic blood film. The cause of the well-known basophilic stippling and the hemolysis is inhibition of pyrimidine 5′ nucleotidase (P5′N) while the hypochromic microcytic anemia and sideroblastic erythropoiesis are due to inhibition of enzymes involved in heme synthesis.

Although lead poisoning is uncommon in most developed countries, hematologists should be alert to the possibility that alternative medications and food supplements containing lead may have been purchased during travel or via the internet. Other causes of lead poisoning include occupational exposure (battery production, foundry work, painting and construction, mining), retained bullets and exposure to lead-based paints in older, poorly maintained houses (sometimes through pica).[2,3] In developing countries, sources of lead are much more widespread than in Western countries and include, in addition to traditional medicines, cosmetics and tooth-cleaning powders, the ongoing use of lead-based paints, water contamination from lead pipes and cisterns, use of lead-containing cookware and lead-glazed crockery and unsafe practices in small factories and mines.[3,4]

Original publication: Bain BJ (2014) Lead poisoning. *Am J Hematol*, **89**, 1141.

Hematology: 101 Morphology Updates, First Edition. Barbara J. Bain.
© 2023 John Wiley & Sons Ltd. Published 2023 by John Wiley & Sons Ltd.

References

1 Saper RB, Phillips RS, Sehgal A, Khouri N, Davis RB, Paquin J *et al.* (2008) Lead, mercury, and arsenic in US- and Indian-manufactured ayurvedic medicines sold via the Internet. *JAMA*, **300**, 915–913.
2 Centers for Disease Control and Prevention (2013) Very high blood lead levels among adults – United States, 2002–2011. *MMWR Morb Mortal Wkly Rep*, **62**, 967–971.
3 Hore P, Ahmed M, Nagin D and Clark N (2014) Intervention model for contaminated consumer products: a multifaceted tool for protecting public health. *Am J Public Health*, **104**, 1377–1383.
4 Pfadenhauer LM, Burns J, Rohwer A and Rehfuess ES (2014) A protocol for a systematic review of the effectiveness of interventions to reduce exposure to lead through consumer products and drinking water. *Syst Rev*, **3**, 36.

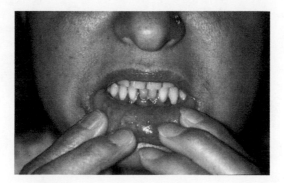

Clinical photograph of the patient showing lead line on gums.

11 Dysplastic neutrophils in an HIV-positive woman

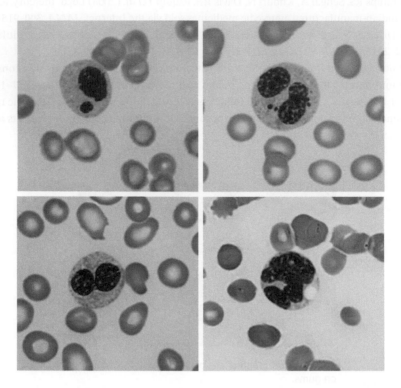

A routine blood count on an 18-year-old HIV-positive African woman showed anemia and thrombocytopenia (Hb 55 g/l, platelet count 10×10^9/l). The blood film showed changes reflecting recurrent infection including increased background staining, rouleaux formation and reactive changes in lymphocytes. In addition, neutrophils showed striking dysplastic features, which included detached nuclear fragments (top images), acquired Pelger–Huët anomaly (bottom left), chromatin clumping, neutrophils with strangely shaped nuclei and a high nucleocytoplasmic ratio (bottom right) and macropolycytes. Dysplastic changes in neutrophils are common in patients with AIDS and can be the feature that suggests the possibility of HIV infection.[1] The presence of detached nuclear fragments in neutrophils is particularly suggestive. The range of changes seen differ from those that are usual in myelodysplastic syndromes. Hypogranularity is less common whereas bizarrely shaped nuclei and a high nucleocytoplasmic ratio in mature cells are more common. Detached nuclear fragments can occur in myelodysplastic syndromes but they are quite uncommon, whereas they are characteristic of HIV infection.

Original publication: Bain BJ (2008) Dysplastic neutrophils in an HIV-positive woman. *Am J Hematol*, **83**, 738.

Reference

1 Bain BJ (1997) The haematological features of HIV infection. *Br J Haematol*, **99**, 1–8.

12   Help with HELLP

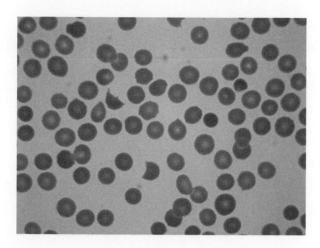

The blood film illustrated is from a 31-year-old woman who presented at 26 weeks' gestation in her first pregnancy with epigastric pain, diarrhea and vomiting. She was found to be hypertensive and edematous with a dipstick test of her urine being strongly positive for protein. Initially her blood count and liver function tests were normal but her hypertension was difficult to control and by 6 days from admission she had developed thrombocytopenia, followed by anemia and abnormal liver function tests. An elective caesarean section was carried out 8 days after admission. Blood tests were at their worst 2 days later: Hb 89 g/l, platelets 25×10^9/l, γ glutamyl transferase 35 iu/l, alanine transaminase 492 iu/l, aspartate transaminase 295 iu/l, alkaline phosphatase 152 iu/l and bilirubin 92 μmol/l. The blood film (image) confirmed the severe thrombocytopenia and showed schistocytes including spheroschistocytes. Recovery occurred thereafter and the baby also survived in good condition.

The HELLP syndrome is characterized by *h*emolysis, *e*levated *l*iver enzymes and *l*ow *p*latelet count. It represents a severe form of pre-eclampsia/eclampsia. The hemolysis is microangiopathic in nature and examination of a blood film is helpful in making the diagnosis.

Original publication: Bain BJ and Riches J (2010) Help with HELLP. *Am J Hematol*, **85**, 70.

13 Neutrophil dysplasia induced by granulocyte colony-stimulating factor

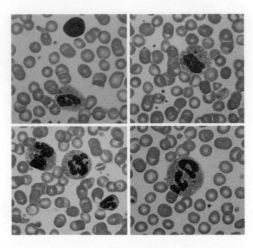

In interpreting dysplastic features and considering a diagnosis of myelodysplastic syndrome (MDS) it is important to be aware of the many other potential causes of dysplasia. One of these is granulocyte colony-stimulating factor (G-CSF), which causes neutrophil dysplasia. The present patient had adult T-cell leukemia/lymphoma and following a course of combination chemotherapy became neutropenic. This led to filgrastim (recombinant human G-CSF) being administered. Neutrophils increased in number and, having been previously cytologically normal, became abnormal (images).

It is not surprising that G-CSF causes neutrophilia, left shift, toxic granulation and Döhle bodies (top left). However, there are other features that would not have been anticipated. One of these is the presence of detached nuclear fragments (top right). Another is the presence of macropolycytes (lower images). The bottom left image shows one neutrophil of normal size and two macropolycytes; one of these has a detached nuclear fragment and the other has a large nucleus of a very abnormal shape. The bottom right image shows another macropolycyte, also with a small detached nuclear fragment; its nucleus is composed of two or more-or-less equal halves joined by a long filament. It is highly probably that this and the other macropolycytes are tetraploid cells.

Detached nuclear fragments were first recognized in a patient taking azathioprine.[1] Subsequently, they were found to be typical of neutrophil dysplasia induced by HIV infection.[2] They are also a feature of G-CSF-induced dysplasia whereas they are quite uncommon in MDS. Macropolycytes are likewise a feature of G-CSF-induced dysplasia but are uncommon in MDS. Conversely, hypogranularity is not a feature of G-CSF-induced dysplasia.

Consideration of the precise nature of any dysplastic features together with an awareness of all the clinical circumstances usually permits a distinction between MDS and other causes of dysplasia.

Original publication: Bain BJ and Nam D (2010) Neutrophil dysplasia induced by granulocyte colony-stimulating factor. *Am J Hematol*, **85**, 354.

References

1 Bain BJ (1989) *Blood Cells; A Practical Guide*. Gower Medical Publishing, London, p. 42.
2 Bain BJ (1997) The haematological features of HIV infection. *Br J Haematol*, **99**, 1–8.

Hematology: 101 Morphology Updates, First Edition. Barbara J. Bain.
© 2023 John Wiley & Sons Ltd. Published 2023 by John Wiley & Sons Ltd.

14 COVID-19 and acute kidney injury

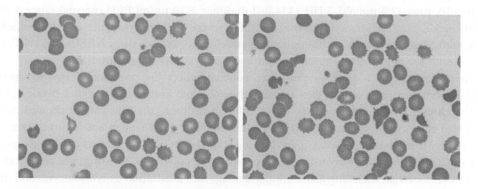

Since the onset of the corona virus disease 2019 (COVID-19) pandemic the spectrum of disease caused by SARS-CoV-2 (severe acute respiratory distress syndrome corona virus 2) has been found to be very wide with a myriad of hematological manifestations. Illustrated here are the peripheral blood features of two patients with acute kidney injury, which is part of this spectrum.

The images above are from a 15-year-old Iraqi patient with β thalassemia major who was on regular transfusion and iron chelation therapy. He presented pericardiac arrest with ventricular tachycardia and acute kidney injury. He was found to have suffered hemorrhage from a duodenal ulcer, requiring emergency surgery and massive transfusion. PCR for SARS-CoV-2 was positive. Postoperatively he required hemofiltration and ventilatory support. His blood count and biochemical tests showed WBC $8.4 \times 10^9/l$, Hb 99 g/l, platelet count $130 \times 10^9/l$, creatinine 125 μmol/l (normal range 50–120), urea 13.2 mmol/l (2.5–7.0), lactate dehydrogenase (LDH) 933 iu/l, D-dimer 8460 ng/ml fibrinogen equivalent units (FEU) (normal range for age 160–390) and serum ferritin 2403 μg/l. Liver function tests were also abnormal. His blood film showed schistocytes and echinocytes (both images above), indicative of microangiopathy and renal insufficiency, and in addition there were occasional pincer or 'mushroom' cells (right image). Further complications during the clinical course included femoral vein thrombosis and seizures due to intracranial microhemorrhages. The patient was treated with dexamethasone and remdesivir and after 19 days in the intensive care unit and a further 13 days of hospitalization was discharged with recovery of renal function to baseline.

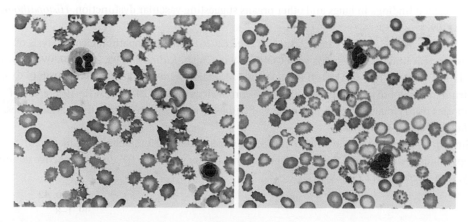

The images above are from a 79-year-old Indian man who presented with fever and respiratory failure. PCR for SARS-CoV-2 was positive. His blood count and biochemistry tests showed WBC 29.5×10^9/l with neutrophilia and monocytosis, Hb 72 g/l, platelet count 28×10^9/l, creatinine 123.8 µmol/l, urea 32.3 mmol/l, LDH 1434 iu/l, D-dimer 2110 ng/ml FEU (normal range 0–500) and ferritin 548.2 µg/l. His blood film confirmed the severe thrombocytopenia and showed schistocytes, marked echinocytosis (including echinocytic elliptocytes) and three NRBCs/100 WBC. The patient's condition deteriorated rapidly and he died within 3 days of presentation.

The peripheral blood features of these two patients reflect the presence of two recognized COVID-19 complications: echinocytes are indicative of acute kidney injury and schistocytes are indicative of microangiopathic damage to red cells. In addition, the first patient showed pincer cells, which have been reported in COVID-19 and suggest possible oxidant-induced damage.[1] Microvascular thrombosis, particularly in the lungs but also in other organs, is a well-recognized feature of COVID-19 and is responsible for the schistocytes observed. The origin of acute kidney injury is multifactorial, including acute tubular damage and infiltration by lymphocytes and macrophages, with the virus having been identified within the glomerular endothelium and in tubular cells. In addition, thrombotic microangiopathy involving the kidney can lead to coagulative necrosis.[2] However, microvascular thrombosis within the kidney does not appear to be common and may not be extensive; microvascular thrombi were observed in only three of 21 cases in one series,[3] in three of 26 cases in another,[4] in one of three cases observed by the authors in a third series[5] and in none of 12 cases in a fourth series.[6] Acute kidney injury in COVID-19 is an adverse prognostic indicator.

Original publication: Crossette-Thambiah C, Hazarika B and Bain BJ (2021) Covid-19 and acute kidney injury. *Am J Hematol*, **96**, 747–748.

References

1 Gérard D, Ben Brahim S, Lesesve JF and Perrin J (2021) Are mushroom-shaped erythrocytes an indicator of COVID-19? *Br J Haematol*, **190**, 230.

2 Jhaveri KD, Meir LR, Chang, Parikh R, Wanchoo R, Barilla-LaBarca ML *et al.* (2020) Thrombotic microangiopathy in a patient with COVID-19. *Kidney Int*, **98**, 509–512.

3 Menter T, Haslbauer JD, Nienhold R, Savic S, Hopfer H, Deigendesch N *et al.* (2020) Postmortem examination of COVID-19 patients reveals diffuse alveolar damage with severe capillary congestion and variegated findings in lungs and other organs suggesting vascular dysfunction. *Histopathology*, **77**, 198–209.

4 Su H, Yang M, Wan C, Yi LX, Tang F, Zhu H-Y *et al.* (2020) Renal histopathological analysis of 26 postmortem findings of patients with COVID-19 in China. *Kidney Int*, **98**, 219–227.

5 Buja LM, Wolf DA, Zhao B, Akkanti B, McDonald M, Lelenwa L *et al.* (2020) The emerging spectrum of cardiopulmonary pathology of the coronavirus disease 2019 (COVID-19): report of 3 autopsies from Houston, Texas, and review of autopsy findings from other United States cities. *Cardiovasc Pathol*, **48**, 107233.

6 Wichmann D, Sperhake JP, Lütgehetmann M, Steurer S, Edler C, Heinemann A *et al.* (2020) Autopsy findings and venous thromboembolism in patients with COVID-19: a prospective cohort study. *Ann Intern Med*, **173**, 268–277.

15 Diagnosis of pyrimidine 5′ nucleotidase deficiency suspected from a blood film

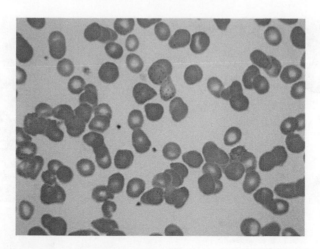

A 21-year-old Kuwaiti female presented with severe anemia. She gave a history that she had been anemic and jaundiced at birth and had needed a blood transfusion. Thereafter she continued to be anemic and had been transfused on a further nine occasions. At the age of 15 years, she had required cholecystectomy for gall stones. The patient was found to have mild hepatomegaly and enlargement of the spleen 23 cm below the left costal margin. A blood count showed Hb 81 g/l and reticulocyte count 13.7%. Bilirubin was 120 µmol/l and lactate dehydrogenase 1136 iu/l. Laboratory facilities to make a precise diagnosis of the hemolytic anemia were not available. Shortly after presentation, the patient suffered blunt trauma to the abdomen and required splenectomy.

Subsequently a pre-splenectomy blood film was reviewed and showed polychromatic macrocytes and prominent basophilic stippling (image), which varied from fine to coarse. The blood film findings, in the context of a congenital hemolytic anemia, suggested a diagnosis of pyrimidine 5′ nucleotidase (P5′N) deficiency and a blood sample was therefore sent from Kuwait to the UK for testing. Deficiency was confirmed: red cell P5′N was 1.0 nmol/h/mg Hb (normal range 9–20) with both parents having reduced activity (mother 4.0, father 6.0 nmol/h/mg Hb), consistent with heterozygosity. At this time, the patient's Hb was 90 g/l and the reticulocyte count 33% (867×10^9/l). Interestingly, a blood film made from the blood that had been couriered from Kuwait to the UK did not show any basophilic stippling, indicating the importance of a fresh blood sample for this morphological observation.

Despite the sophisticated modern methods available for making a precise diagnosis in patients with hemolytic anemia, careful examination of a blood film can still provide a significant clue to the correct diagnosis. This case also illustrates the value of retaining blood films, particularly when there is an unresolved diagnostic problem.

Original publication: Al-Jafar HA, Layton DM, Robertson L, Escuredo E and Bain BJ (2013) Diagnosis of pyrimidine 5′-nucleotidase deficiency suspected from a blood film. *Am J Hematol*, **88**, 1089.

Hematology: 101 Morphology Updates, First Edition. Barbara J. Bain.
© 2023 John Wiley & Sons Ltd. Published 2023 by John Wiley & Sons Ltd.

16 Bone marrow aspirate in Chédiak–Higashi syndrome

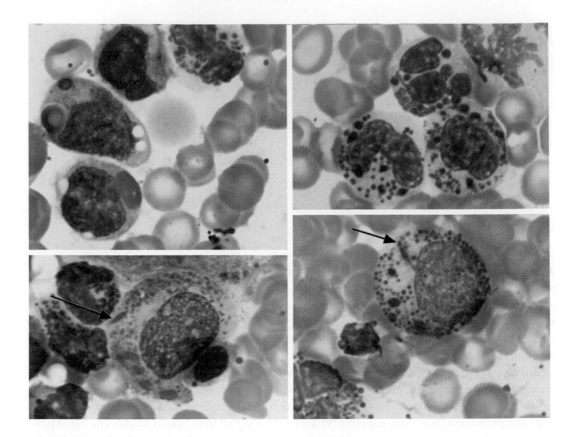

A 10-month-old Iraqi male presented with meningoencephalitis, treated with antibiotics. His blood count showed WBC 2.5 × 10⁹/l, neutrophil count 0.08 × 10⁹/l, Hb 96 g/l and platelet count 160 × 10⁹/l. The severe neutropenia led to a bone marrow aspirate being performed. This revealed the diagnosis of Chédiak–Higashi syndrome. The child's skin was normally pigmented and the diagnosis had not previously been suspected.

The inclusions of Chédiak–Higashi syndrome sometimes have staining characteristics appropriate to a lineage but sometimes they stain abnormally. In these images of a Romanowsky-stained bone marrow film, eosinophil granules are giant but retain their eosinophilic characteristics whereas many of the giant neutrophil granules stain deep purple. In addition to globular inclusions, some cells contain crystals (arrows).

Original publication: Abdulsalam AH, Sabeeh N and Bain BJ (2011) Bone marrow aspirate in Chédiak-Higashi syndrome. *Am J Hematol*, **87**, 100.

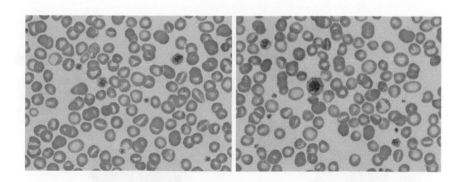

17 Phytosterolemia

A 12-year-old Iranian boy was referred because of anemia that was refractory to iron therapy despite a low serum iron. His parents were first cousins and he was known to have β thalassemia heterozygosity. He was on the ninth centile for height and weight. His blood count showed Hb 88 g/l with MCV 56.8 fl, MCH 17.6 pg, MCHC 309 g/l and platelet count 209 × 10⁹/l. His anemia had the features of anemia of chronic disease with a low serum iron, low transferrin, low transferrin saturation and ferritin of 375 µmol/l. A blood film showed microcytosis and basophilic stippling, consistent with the known β thalassemia trait. However, there were also features suggesting a diagnosis of phytosterolemia (also known as sitosterolemia). Specifically, there was a combination of stomatocytes and large platelets (images). Lactate dehydrogenase was increased to 249 iu/l and on two occasions the reticulocytes were 156 and 191 × 10⁹/l, consistent with hemolysis. The diagnosis was confirmed by measurement of the plasma lipids. There was an increase of cholesterol and triglycerides but in addition plant sterols were strikingly increased: campesterol 1008 µmol/l (normal range 0–56), stigmasterol 31 µmol/l (normal zero), sitosterol 1324 µmol/l (0–41), fucosterol 187 µmol/l (zero) and stigmastanol 124 µmol/l (zero). Investigation of the patient's brother showed him to have the same condition. The patient was commenced on therapy with ezetimibe, a sterol pump inhibitor that reduces absorption of plant sterols.

Phytosterolemia is an autosomal recessive condition resulting from mutation in either the *ABCG5* or the *ABCG8* genes, in which there is excessive absorption of dietary sterols including plant sterols. As a result there is formation of xanthomas and premature atherosclerosis. The condition appears to be rare but it is likely that it is underdiagnosed. Examination of a blood film, with recognition of the significance of the particular combination of large platelets and stomatocytosis, is critical in making the diagnosis. Often the platelet count is reduced. Diagnosis is important because of the premature vascular disease and because patients who are misdiagnosed have sometimes been treated with corticosteroids or splenectomy.[1] Correct diagnosis has become particularly important with the availability of specific treatment.

Original publication: Bain BJ and Chakravorty S (2016) Phytosterolemia. *Am J Hematol*, **91**, 643.

Reference

1 Wang Z, Cao L, Su Y, Wang G, Wang R, Yu Z *et al*. (2014) Specific macrothrombocytopenia/hemolytic anemia associated with sitosterolemia. *Am J Hematol*, **89**, 320–324.

Hematology: 101 Morphology Updates, First Edition. Barbara J. Bain.
© 2023 John Wiley & Sons Ltd. Published 2023 by John Wiley & Sons Ltd.

18 Pseudo-Chédiak–Higashi inclusions together with Auer rods in acute myeloid leukemia

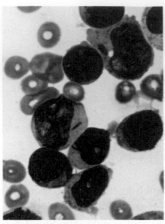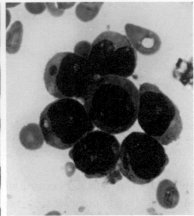

An 8-year-old Iraqi male presented with fever and pallor of 2 months' duration. There was no lymphadenopathy, splenomegaly or hepatomegaly. A blood count showed WBC 1.5×10^9/l, Hb 69 g/l and platelet count 217×10^9/l. Examination of a bone marrow aspirate showed acute myeloid leukemia (AML) with little maturation; blast cells were 85% of nucleated cells, and the case was classified as FAB (French–American–British) M1 AML. An unusual feature was the presence not only of Auer rods (left image) but also of pseudo-Chédiak–Higashi inclusions, which on a Romanowsky stain varied from pink to deep purple (right). In addition to typical thin Auer rods there were also thick rod-shaped inclusions (left and right).

Pseudo-Chédiak–Higashi inclusions are observed occasionally in AML[1] (including acute promyelocytic leukemia) and also, rarely, in refractory anemia with excess of blasts.[2] The inclusions are formed by fusion of azurophilic (primary) granules and show myeloperoxidase activity and Sudan black B staining. On ultrastructural examination they are heterogeneous and, in contrast to the inclusions seen in the inherited anomaly, contain rod-shaped structures showing periodicity.[1] The diagnostic significance of these inclusions is likely to be similar to that of Auer rods, indicating either AML or a high grade myelodysplastic syndrome.

Original publication: Abdulsalam AH, Sabeeh N and Bain BJ (2011) Pseudo-Chédiak–Higashi inclusions together with Auer rods in acute myeloid leukemia. *Am J Hematol*, **86**, 602.

References

1 Tulliez M, Vernant JP, Breton-Gorius J, Imbert M and Sultan C (1979) Pseudo-Chediak–Higashi anomaly in a case of acute myeloid leukemia: electron microscopic studies. *Blood*, **54**, 863–871.
2 Bain BJ (2010) *Leukaemia Diagnosis*, 6th edn. Wiley-Blackwell, Oxford, pp. 112–113.

19 Botryoid nuclei resulting from cocaine abuse

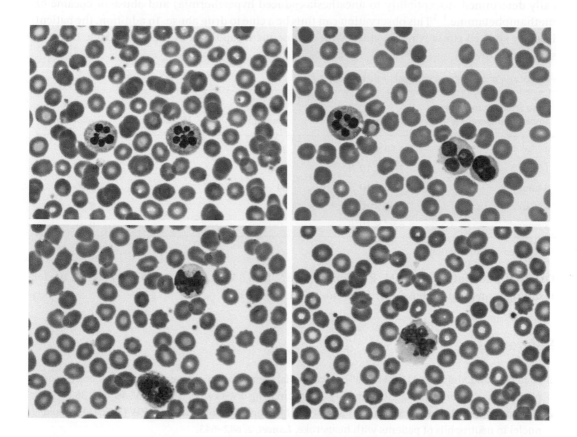

A blood count on a 32-year-old, seriously ill patient presenting to the emergency department, showed lymphocytosis (lymphocyte count 5.8×10^9/l). A blood film was therefore examined. This showed numerous neutrophils with botryoid nuclei (top images), lymphocytes with multilobated nuclei (top right) and monocytes with butterfly- and fan-shaped nuclei (bottom right); basophils and eosinophils (bottom left, which also shows a lobulated large granular lymphocyte) also had botryoid nuclei. Two hours later a blood count from the patient, now in the intensive care unit, showed that the lymphocyte count had fallen to 0.50×10^9/l and surprisingly that the nuclear aberrations were no longer present. The history was then obtained that the patient had multiorgan failure and had presented with a temperature of 41.9°C. Cocaine abuse was suspected and measures were taken to lower the body temperature. Subsequently, urinary breakdown products of drugs of abuse (opiates, cocaine and cannabinoids) were detected.

We were interested in the rapid disappearance of the abnormal leukocytes and sought to explain this by an *in vitro* experiment. A K_2-EDTA-anticoagulated blood sample was incubated at 41.9°C for 1 hour. A blood count and films were then performed and these were repeated over the next 8 hours with storage at ambient temperature (table below). Reversion to normal cytology was demonstrated. This nuclear abnormality is thus reversible within certain limits.

Hematology: 101 Morphology Updates, First Edition. Barbara J. Bain.
© 2023 John Wiley & Sons Ltd. Published 2023 by John Wiley & Sons Ltd.

Botryoid nuclei are a feature of burns and of hyperthermia of diverse origins including heatstroke, brainstem hemorrhage or ischemia, encephalitis, malignant hyperthermia (genetically determined susceptibility to anesthesia-induced hyperthermia) and abuse of cocaine or methamphetamine.[1–3] This observation can thus be a clue to drug abuse. In addition, the patient initially showed stress lymphocytosis with a subsequent rapid fall of the lymphocyte count to subnormal levels.

Percentage of leukocytes with botryoid nuclei

Time after returning to ambient temperature	White cell count	Neutrophils	Monocytes	Lymphocytes
0	$7.10 \times 10^9/l$	21	98	23
1 hour	$7.09 \times 10^9/l$	56	99	43
3 hours	$7.06 \times 10^9/l$	34	56	21
4 hours	$7.07 \times 10^9/l$	12	27	9
8 hours*	$6.92 \times 10^9/l$	0	2	0

* Increased apoptotic forms

Original publication: Fumi M, Pancione Y, Sale S, Rocco V and Bain BJ (2017) Botryoid nuclei resulting from cocaine abuse. *Am J Hematol*, **92**, 1260–1261.

References

1 Neftel KA and Muller OM (1981) Heat-induced radial segmentation of leucocyte nuclei: a non-specific phenomenon accompanying inflammatory and necrotizing diseases *Br J Haematol*, **48**, 377–382.

2 Hernandez JA, Aldred SW, Bruce JR, Vanatta PR, Mattingly TL and Sheehan WW (1980) "Botryoid" nuclei in neutrophils of patients with heatstroke. *Lancet*, **2**, 642–643.

3 Im DD, Ho CH and Chang RY (2015) Hypersegmented neutrophils in an adolescent male with heatstroke. *J Pediatr Hematol Oncol*, **37**, 488.

20 Infantile pyknocytosis

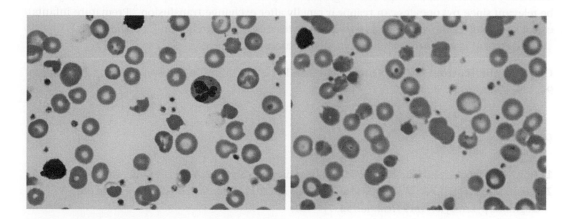

A newborn baby boy, born at term to consanguineous parents of Moroccan origin, presented aged 2 weeks with jaundice. He was breastfeeding well, was gaining weight and was not lethargic. His blood count showed WBC 22.3×10^9/l, Hb 63 g/l, MCV 102 fl and platelet count 596×10^9/l. Total bilirubin was 396 μmol/l and reticulocytes were 404×10^9/l (20.5%). He had been jaundiced at 48 hours of age and was treated with double phototherapy for 24 hours; a point-of-care capillary Hb had been 181 g/l at that time. There was RhD incompatibility; his mother was a primigravida and had received appropriate anti-D prophylaxis and had had negative antibody screens throughout. The direct antiglobulin test was negative.

His peripheral blood film (images) showed keratocytes, irregularly contracted cells, blister cells, polychromasia and some Howell–Jolly bodies, consistent with oxidative hemolysis. A Heinz body screen was positive. There was no history of exposure to any oxidizing agents in either the mother or baby. Pretransfusion investigations included glucose-6-phosphate dehydrogenase (G6PD) assay, which was 18.8 u/g Hb (pediatric normal range 10–15). (G6PD level in the mother was normal at 10.1 u/g Hb.) Pyruvate kinase was high at 18.9 u/g Hb (adult normal range 6.2–14.2). High performance liquid chromatography was normal for age: hemoglobin A 54%, hemoglobin A_2 1.1% and hemoglobin F 46.2%. Hemoglobin mass spectrometry was normal. Glutathione peroxidase level was 9.2 u/g Hb (adult normal range 13.0–30.0). Selenium was 0.77 μmol/L (adult normal range 0.25–0.6). In view of the morphological findings and the reduced glutathione peroxidase, a diagnosis of infantile pyknocytosis was made. Subsequent analysis of the *GPX1* gene disclosed no mutation.

The baby was transfused red cells and discharged home. Follow-up blood tests showed resolving hemolysis; at 2 months of age Hb was 104 g/l and reticulocytes 81×10^9/l with no further need for transfusion.

Infantile pyknocytosis is a poorly understood condition, first described by Tuffy *et al.* in 1959.[1,2] These authors commented on 'markedly distorted and contracted red cells', which they also called 'burr cells'. Acanthocytes and Howell–Jolly bodies can also be present, reflecting splenic overload. The 'pyknocytes' are keratocytes and other dense deformed cells. They appear as hyperdense cells on automated instrument scatter plots. The diagnosis rests on the characteristic blood film features of oxidant damage, with other conditions such as G6PD deficiency and exposure to an exogenous oxidant being excluded. As in this patient, there can be a transient neonatal deficiency of glutathione peroxidase, in the absence of any mutation in the gene. Other cases are attributed to deficiency of selenium, a component of glutathione peroxidase. Extracorpuscular factors may also be relevant since pyknocytes can reappear rapidly after exchange transfusion.[3] Premature infants can also develop hemolytic anemia with pyknocytes as a result of vitamin E deficiency.[4]

Management is supportive since spontaneous resolution occurs.

Original publication: Rees C, Lund K and Bain BJ (2019) Infantile pyknocytosis. *Am J Hematol*, **94**, 489–490.

References

1 Tuffy P, Brown AK and Zuelzer WW (1959) Infantile pyknocytosis. *Am J Dis Child*, **98**, 227–241.
2 Nassin ML, Vergilio J-A, Heeney MM and LaBelle JL (2017) Neonatal anemia: revisiting the enigmatic pyknocyte. *Am J Hematol*, **92**, 717–721.
3 Ackerman BD (1969) Infantile pyknocytosis in Mexican-American infants. *Am J Dis Child*, **117**, 417–423.
4 Oski FA and Barness LA (1967) Vitamin E deficiency: a previously unrecognized cause of hemolytic anemia in the premature infant. *J Pediatr*, **70**, 211–220.

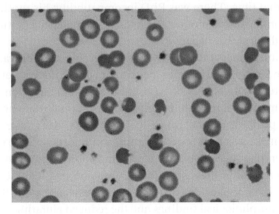

A further image showing irregularly contracted cells, which are acanthocytic.

21 Splenic rupture in cytomegalovirus infection

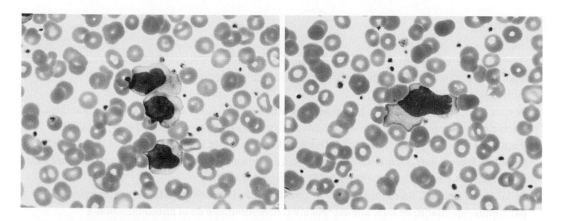

A 37-year-old man presented to the emergency department with acute abdominal pain and symptomatic hypotension, with a blood pressure on admission of 60/33 mmHg. His blood count showed Hb 132 g/l, WBC 5.7×10^9/l, platelet count 222×10^9/l, neutrophils 3.0×10^9/l and lymphocytes 2.1×10^9/l. On examination, he had tenderness to palpation of his left upper quadrant and was thought to have suffered a spontaneous splenic rupture. Computed tomography showed a large perisplenic high attenuation collection. He had an urgent laparotomy and splenectomy with an estimated blood loss of 4.5 litres.

His blood count 10 days after surgery showed Hb 92 g/l, WBC 11.4×10^9/l, lymphocyte count 8.1×10^9/l and platelet count 1068×10^9/l, and a blood film was made. This showed a significant thrombocytosis with platelet anisocytosis. Howell–Jolly bodies were also seen. However, in addition, there were atypical lymphocytes, which were large and irregular in shape with reduced condensation of chromatin and voluminous basophilic cytoplasm, sometimes containing vacuoles or azurophilic granules (images). No blood film had been made on presentation since the blood count was normal and there was no instrument flag suggesting atypical lymphocytes. Viral serology was positive for cytomegalovirus (CMV) IgM and IgG antibodies, indicating the patient had current CMV infection. CMV DNA PCR was positive and a viral load of 10,600 copies/ml was detected. The patient was treated with a 5-day course of aciclovir 400 mg five times a day.

Spontaneous (atraumatic) splenic rupture is rare and is most commonly caused by an underlying hematologic malignancy or infection. Rarer causes include: hematological disorders causing an increased bleeding tendency such as hemophilia or autoimmune thrombocytopenia; splenomegaly associated with hemolytic anemia; inflammatory causes such as rheumatoid arthritis and polyarteritis nodosa; primary amyloidosis; iatrogenic causes including use of anticoagulants or granulocyte colony-stimulating factor (including cases in healthy stem cell donors); and primary splenic disorders such as splenic cyst, malignancy or infarction. Infectious mononucleosis is a well-recognized cause but there have also been previous reports of cases due to primary CMV infection.[1–15] In one patient, there was not only spontaneous splenic rupture but also very severe virus-related thrombocytopenia underlying the life-threatening hemorrhage.[15] Occasional patients have been managed conservatively[4,10] but most have required emergency splenectomy.

Hematology: 101 Morphology Updates, First Edition. Barbara J. Bain.
© 2023 John Wiley & Sons Ltd. Published 2023 by John Wiley & Sons Ltd.

Original publication: Farnell R, Siow W and Bain BJ (2019) Splenic rupture in cytomegalovirus infection. *Am J Hematol*, **94**, 828–829.

References

1 Horwitz CA, Henle W, Henle G, Snover D, Rudnick H, Balfour H *et al.* (1986) Clinical and laboratory evaluation of cytomegalovirus-induced mononucleosis in previously healthy individuals. *Medicine*, **3**, 124–134.

2 Rogues AM, Dupon M, Cales V, Malou M, Paty MC, Le Bail B and Lacut JY (1994) Spontaneous splenic rupture: an uncommon complication of cytomegalovirus infection. *J Infect*, **29**, 83–85.

3 Ragnaud J, Morlat P, Gin H, Dupon M, Delafaye C, du Pasquier P and Aubertin J (1994) Aspects cliniques, biologiques et évolutifs de l'infection à cytomégalovirus chez le sujet immunocompétent: à propos de 34 patients hospitalisés. *Rev Med Interne*, **15**, 13–18.

4 Bellaïche G, Habib E, Baledent F, Nouts A, Lusina D, Ley G and Slama JL (1998) Hemoperitoine par rupture spontanee de rate: une complication exceptionnelle de la primo-infection a CMV. *Gastroenterol Clin Biol*, **22**, 107–108.

5 Alliot C, Beets C, Besson M and Derolland P (2001) Spontaneous splenic rupture associated with CMV infection: report of a case and review. *Scand J Infect Dis*, **33**, 875–877.

6 Duarte PJ, Echavarria M, Paparatto A and Cacchione R (2003) Ruptura espontánea de bazo asociada a infección activa por citomegalovirus. *Medicina*, **63**, 46–48.

7 Gogone S, Praticò S, Di Pietro N, Melita G, Sanò M, De Luca M *et al.* (2005) Rottura spontanea di milza in presenza di infezione citomegalica. Descrizione di un caso clinico. *G Chir*, **26**, 95–99.

8 Amathieu R, Tual L, Rouaghe S, Stirnemann J, Fain O and Dhanneur G (2007) Rupture spontanée de la rate au cours d'une infection à cytomégalovirus: cas clinique et revue de la littérature. *Ann Fr Anesth Reanim*, **26**, 674–676.

9 Maillard N, Koenig M, Pillet S, Cuilleron M and Cathébra P (2007) Spontaneous splenic rupture in primary cytomegalovirus infection. *Presse Med*, **36**, 874–877.

10 Lianos G, Ignatiadou E, Bali C, Harissis H and Katsios C (2012) Successful nonoperative management of spontaneous splenic hematoma and hemoperitoneum due to CMV infection. *Case Rep Gastrointest Med*, **2012**, 328474.

11 Maria V, Saad AM and Fardellas I (2013) Spontaneous spleen rupture in a teenager: an uncommon cause of acute abdomen. *Case Rep Med*, **2013**, 675372.

12 de Havenon A, Davis G and Hoesch R (2014) Splenic rupture associated with primary CMV infection, AMSAN, and IVIG. *J Neuroimmunol*, **272**, 103–105.

13 Vidarsdottir H, Bottiger B and Palsson B (2014) Spontaneous splenic rupture and multiple lung embolisms due to cytomegalovirus infection: a case report and review of the literature. *Int J Infect Dis*, **21**, 13–14.

14 Roche M, Maloku M and Abdel-Aziz TE (2014) An unusual diagnosis of splenic rupture. *BMJ Case Rep*, **2014**, bcr2014204891.

15 Glesner MK, Madsen KR, Nielsen JM and Posth S (2015) Successful emergency splenectomy during cardiac arrest due to cytomegalovirus-induced atraumatic splenic rupture. *BMJ Case Rep*, **2015**, bcr2014208094.

22 A new diagnosis of monoclonal B-cell lymphocytosis with cytoplasmic inclusions in a patient with COVID-19

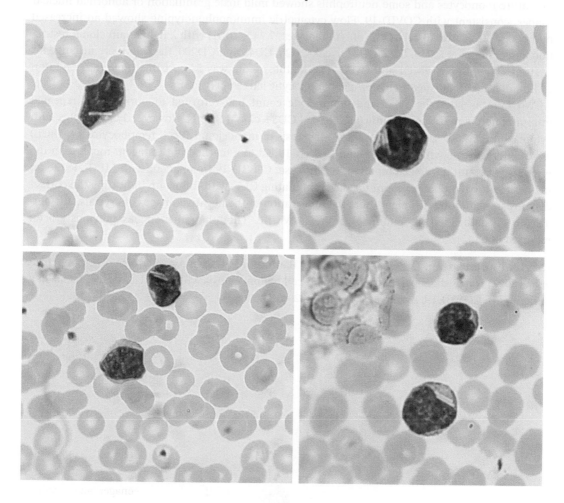

A 75-year-old man with a history of chronic ischemic heart disease with a previously normal blood count, presented to the emergency department with fever and tachycardia. There was no hepatosplenomegaly or lymphadenopathy. An electrocardiogram showed left bundle branch block. Oxygen saturation was 95%. Because of the fever the patient was tested for severe acute respiratory distress syndrome corona virus 2 (SARS-CoV-2); the RT-PCR assay result was positive. A diagnosis of corona virus disease 2019 (COVID-19) was made.

The patient's blood count showed WBC 10.46×10^9/l, lymphocytes 4.51×10^9/l, Hb 129 g/l and platelet count 233×10^9/l. D-dimer was 659 µg/l (normal range, cut off 500 µg/l) and interleukin 6 (IL-6) was 76.3 pg/ml (normal range <6.4). A computed tomography scan of the chest showed bilateral interstitial infiltrates associated with multiple enlarged mediastinal lymph nodes. Following a rapid and unexpected increase of the WBC to 17.49×10^9/l and of the lymphocyte count to 8.37×10^9/l over the next 48 hours, a blood film and immunophenotyping were performed. The film showed small and medium-sized lymphocytes, with a variable N:C ratio and

Hematology: 101 Morphology Updates, First Edition. Barbara J. Bain.
© 2023 John Wiley & Sons Ltd. Published 2023 by John Wiley & Sons Ltd.

moderately basophilic cytoplasm. Smear cells were present. A quarter of the lymphocytes showed the negative images of one to three rod-shaped crystals (average two per cell). There were some immature monocytes and some neutrophils showed mild toxic granulation or abnormal nuclear shapes, consistent with COVID-19. Flow cytometric immunophenotyping showed an increased number of circulating B cells (93% of lymphocytes, $7.78 \times 10^9/l$) with λ light chain clonal restriction and expressing CD19, CD5, CD23, weak CD20, CD43 and CD200; the cells were negative for CD10, CD79b, CD81, FMC7 and CD38. The Matutes score was 5/5. At this stage the condition could not be distinguished from chronic lymphocytic leukemia (CLL).

Two months later the WBC and lymphocyte count had returned to normal and repeated immunophenotyping showed only $0.63 \times 10^9/l$ CD5+ clonal B cells. However, lymphocytes with cytoplasmic crystals were still present. A diagnosis of monoclonal B-cell lymphocytosis (MBCL) was made. Patients with CLL in whom COVID-19 led to a marked but transient increase in the lymphocyte count have been reported. In this case, COVID-19 in a patient with MBCL led to an increase in the lymphocyte count to a level simulating CLL, with follow-up indicating the correct diagnosis. In addition, we report here the observation of the negative images of crystals, attributable to crystallization of immunoglobulin, in MCBL. This phenomenon has previously been reported in CLL.

Original publication: Lanza L, Koroveshi B, Barducchi F, Lorenzo A, Venturino E, Cappelli E *et al.* (2022) A new diagnosis of monoclonal B-cell lymphocytosis with cytoplasmic inclusions in a patient with COVID-19. *Am J Hematol*, **97**, 1372–1373.

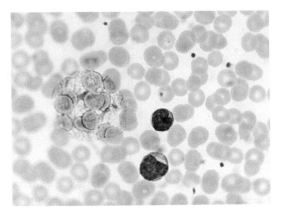

A further image with two lymphocytes, one of which shows the negative image of two crystals.

23 Pseudoplatelets and apoptosis in Burkitt lymphoma

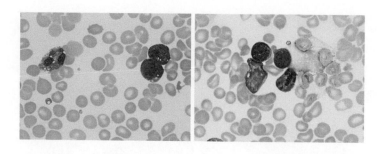

Pseudoplatelets are particles of similar size to platelets that are counted as platelets by automated instruments. They are a known feature of acute myeloid and acute lymphoblastic leukemia, being reported in 30% and 18% of cases, respectively, in a large series of patients.[1] They have also been reported in occasional cases of hairy cell leukemia[2] and lymphocytic lymphoma.[3] Pseudoplatelets can also be a prominent feature of the leukemic phase of Burkitt lymphoma. The images shown are from a 57-year-old HIV-positive African man. His blood count showed WBC 55×10^9/l, Hb 96 g/l and 'platelet count' 39×10^9/l. A blood film showed that the true platelet count was much lower, the factitious elevation being due to the presence of pseudoplatelets. These were easily recognizable as fragments of the cytoplasm of lymphoma cells since they were basophilic and often contained vacuoles. The left image shows two Burkitt lymphoma cells and three pseudoplatelets, one with a prominent vacuole. In the presence of pseudoplatelets, impedance and light scattering counts are erroneous but a correct count can be obtained on instruments that can perform an immunological count. Alternatively, the ratio of platelets to pseudoplatelets can be determined from the blood film, and the count can then be corrected. Recognizing this phenomenon is important since dangerous thrombocytopenia may otherwise be missed.

Another common feature of Burkitt lymphoma is a high rate of apoptosis, which may be reflected in the blood film. The right image shows two intact lymphoma cells, two apoptotic cells, a pseudoplatelet and a smear cell. It may be postulated that the presence of numerous pseudoplatelets is linked to the presence of fragile apoptotic cells.

Original publication: Bain BJ (2019) Pseudoplatelets and apoptosis in Burkitt lymphoma. *Am J Hematol*, **94**, 1168–1169.

References

1 van der Meer W, MacKenzie MA, Dinnissen JWB and de Keijzer MH (2003) Pseudoplatelets: a retrospective study of their incidence and interference with platelet counting. *J Clin Pathol*, **56**, 772–774.
2 Ballard HS and Sidhu G (1981) Cytoplasmic fragments causing spurious platelet counts in hairy cell leukemia: ultrastructural characterization. *Arch Intern Med*, **141**, 942–944.
3 Stass SA, Holloway ML, Peterson V, Creegan WJ, Gallivan M and Schumacher HR (1979) Cytoplasmic fragments causing spurious platelet counts in the leukemic phase of poorly differentiated lymphocytic lymphoma. *Am J Clin Pathol*, **71**, 125–128.

Hematology: 101 Morphology Updates, First Edition. Barbara J. Bain.
© 2023 John Wiley & Sons Ltd. Published 2023 by John Wiley & Sons Ltd.

24 What is a promonocyte?

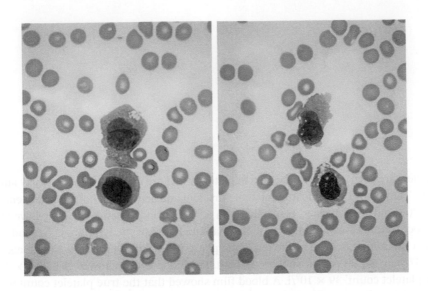

The French–American–British (FAB) cooperative group, in their seminal paper of 1976, described a promonocyte as follows: "This cell is similar to the monoblast but has a large nucleus with a cerebriform appearance; nucleoli may be present, but the cytoplasm is less basophilic, has a greyish ground-glass appearance and fine azurophilic granules are often scattered throughout."[1] They did not advise counting promonocytes with blast cells when distinguishing between acute leukemia and a myelodysplastic syndrome. In 2008, the World Health Organization (WHO) expert group described promonocytes as follows: "promonocytes have a delicately convoluted, folded or grooved nucleus with finely dispersed chromatin, a small indistinct or absent nucleolus, and finely granulated cytoplasm."[2] The two descriptions have some similarities and yet the WHO group advised that monoblasts and promonocytes be summated when estimating the blast count to establish a diagnosis of acute leukemia. They emphasized the importance of distinguishing promonocytes from "more mature but abnormal leukemic monocytes . . . which have more clumped chromatin . . . variably indented, folded nuclei and grey cytoplasm with more abundant lilac-coloured granules."

  These images are from a 70-year-old woman who presented with septicemia and a purpuric rash. Her blood count showed WBC 36×10^9/l, Hb 89 g/l and platelet count 15×10^9/l. A diagnosis of acute monocytic leukemia was made on the basis of morphology and cytochemical stains of peripheral blood and bone marrow. The left image compares a monoblast and a promonocyte. The monoblast (below) has a large nucleolus and a regular oval nucleus whereas the promonocyte (above) has a notch that leads into a groove crossing the nucleus and has smaller, less distinct nucleoli and more obvious granules. The right image compares a promonocyte with an immature abnormal monocyte. The promonocyte (above) has a nucleolus and two superimposed nuclear lobes while the abnormal immature monocyte (below) is showing some definite chromatin condensation; avoiding classifying this latter cell as a promonocyte helps to avoid a misdiagnosis of chronic myelomonocytic leukemia as acute leukemia. It is the finely dispersed chromatin of the promonocyte that is the most important feature in distinguishing it from an immature monocyte.

Hematology: 101 Morphology Updates, First Edition. Barbara J. Bain.
© 2023 John Wiley & Sons Ltd. Published 2023 by John Wiley & Sons Ltd.

Original publication: Bain BJ (2013) What is a promonocyte? *Am J Hematol*, **88**, 919.

References

1 Bennett JM, Catovsky D, Daniel MT, Flandrin G, Galton DAG, Gralnick HR and Sultan C (1976) Proposals for the classification of the acute leukaemias (FAB cooperative group). *Br J Haematol*, **33**, 451–458.

2 Vardiman JW, Brunning RD, Arber DA, Le Beau MM, Porwit A, Tefferi A *et al.* (2008) Introduction and overview of the classification of the myeloid neoplasm. In: Swerdlow SH, Campo E, Lahhir NL, Jaffe ES, Pileri SA, Stein H *et al.* (eds). *WHO Classification of Tumours of Haematopoietic and Lymphoid Tissues.* IARC, Lyon, pp. 17–30.

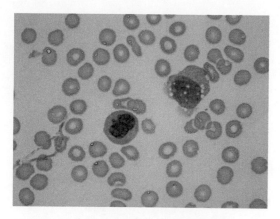

A further image showing a promonocyte with delicate chromatin and overlapping nuclear lobes (top) and an immature monocyte (bottom).

25 Persistent neonatal jaundice resulting from hereditary pyropoikilocytosis

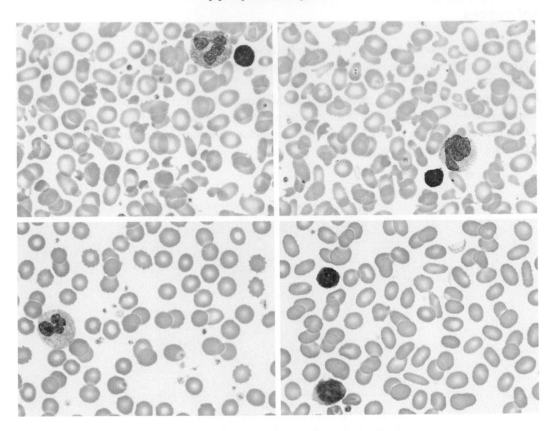

A female infant, weighing 2.39 kg, was born at 35 weeks' gestation, to a mother of Eritrean and a father of Ghanaian ancestry. The neonate had feeding difficulties and was noted to be jaundiced at 36 hours of age. Initial treatment was phototherapy, intravenous fluids and antibiotics for suspected sepsis. Improvement occurred over the next few days; however, it was noted that the bilirubin declined at a slower than anticipated rate. Investigations for neonatal jaundice showed no evidence of alloimmune hemolysis and normal glucose-6-phosphate dehydrogenase. High performance liquid chromatography (HPLC) showed hemoglobins A and F. On day 5 of life the Hb was 133 g/l. The blood film (top images) showed marked anisopoikilocytosis, most prominently ovalocytes, elliptocytes and schistocytes, with occasional spherocytes and nucleated red blood cells. There was polychromasia and the reticulocyte count was 175 and 125×10^9/l on days 4 and 5, respectively. The MCV was low for age at 85 fl, reflecting the large number of schistocytes and small damaged cells.

Given the blood film findings, there was a high suspicion of hereditary pyropoikilocytosis. The neonate was commenced on folic acid and parental blood films were assessed. The mother had an unremarkable blood film (bottom left) while the blood film of the father demonstrated significant elliptocytosis and ovalocytes, indicative of hereditary elliptocytosis (bottom right). A red cell gene panel on the neonate showed two variants: (i) a heterozygous *SPTA1* c.460_462dup; p. (Leu155dup)

Hematology: 101 Morphology Updates, First Edition. Barbara J. Bain.
© 2023 John Wiley & Sons Ltd. Published 2023 by John Wiley & Sons Ltd.

pathogenic variant; and (ii) a heterozygous *SPTA1* c.[5572C>G:6531-12C>T]; p. (Leu1858Val) low expression allele – also known as α spectrin[LELY]. Parental studies showed that the first of these was inherited from the father and the second from the mother. The baby remained well and was discharged home on day 14 of life. However, by day 19 the Hb had fallen to 78 g/l, and a blood transfusion was given.

Hereditary pyropoikilocytosis most often results from coinheritance of a pathogenic variant of *SPTA1*, the gene encoding α spectrin, and a common polymorphism, α spectrin[LELY] (low expression LYon allele), which is present in 20–30% of many populations.[1] Other cases result from homozygosity for a pathogenic *SPTA1* mutation or, less often, homozygosity for a pathogenic mutation in *SPTB*, encoding β spectrin, or coinheritance of mutant *SPTA1* and *SPTB*. The severity of the condition is variable with blood transfusion sometimes being needed and, rarely, splenectomy. The designation of this condition reflects the thermal sensitivity of the cells (*pyro* = fire) with the blood film showing some similarities to that of thermal burns or accidental heating of a blood specimen on its way to the laboratory. Diagnosis starts with morphology but increasingly it is confirmed by next generation sequencing. The differential diagnosis includes hereditary elliptocytosis with transient neonatal poikilocytosis, the differentiation being based on clinical follow-up and genetic studies.

Original publication: Sivaguru G, Simini G and Bain BJ (2022) Persistent neonatal jaundice resulting from hereditary pyropoikilocytosis. *Am J Hematol*, **97**, 506–507.

Reference

1 Nissa O, Chonatb S, Dagaonkarc N, Almansoon MO, Kerr K, Togers ZR *et al.* (2016) Genotype-phenotype correlations in hereditary elliptocytosis and hereditary pyropoikilocytosis. *Blood Cells Mol Dis*, **61**, 4–9.

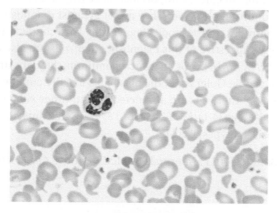

A further image from the neonate showing dense irregular cells and occasional elliptocytes.

26 Auer rods or McCrae rods?

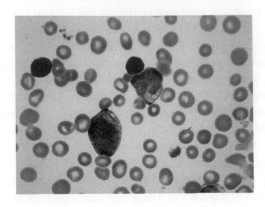

Auer rods are rod-shaped crystalline structures derived from the primary granules of myeloid cells. They are most characteristic of acute myeloid leukemia (AML) but are observed also in high grade myelodysplastic syndromes and myelodysplastic/myeloproliferative neoplasms (myelodysplastic syndrome with excess blasts and chronic myelomonocytic leukemia 2, according to the 2016 World Health Organization classification). These cytoplasmic inclusions are named for John Auer (1875–1948) who described and illustrated them in 1906[1] in a patient admitted to the Johns Hopkins Hospital under the care of William Osler. Auer acknowledged that the observations in this patient had been mentioned previously, in the preceding year, in an article by Thomas McCrae who was also acknowledged for the clinical notes he provided for Auer's paper.[2,3] McCrae in his paper had mentioned the forthcoming more detailed description by Auer.[2] It has been suggested that the designation McCrae–Auer rods would be appropriate.[3] Interestingly, both McCrae and Auer thought that the cells in which they observed the inclusions were lymphoblasts.

Auer rods are of considerable diagnostic importance since they indicate both the lineage and the neoplastic nature of the condition observed. They are of greatest diagnostic significance in acute promyelocytic leukemia. Bundles of Auer rods ('faggot cells') are characteristic and, although not totally specific, are strongly suggestive of this diagnosis. Auer rods are also characteristic of AML with t(8;21)(q22;q22.1)/*RUNX1::RUNX1T1* in which condition they are unusually long (image) and are often located in a visible Golgi zone; the blast cells usually contain only a single rod, but occasionally two are present. Auer rods are also detectable in other subtypes of AML, including AML associated with t(6;9)(q23;q34.1)/*KMT2A::MLLT3*. They can thus be a feature of both good prognosis and bad prognosis subtypes of AML.

Original publication: Bain BJ (2011) Auer rods or McCrae rods? *Am J Hematol*, **86**, 689.

References

1 Auer J (1906) Some hitherto undescribed structures found in the large lymphocytes of a case of acute leukemia. *Am J Med Sci*, **131**, 1002–1015.
2 McCrae T (1905) Acute lymphatic leukaemia, with the report of five cases. *BMJ*, 1, 404–408.
3 Seymour JF (2006) 101 years of McCrae's (not Auer's) rods. *Br J Haematol*, **133**, 690.

Hematology: 101 Morphology Updates, First Edition. Barbara J. Bain.
© 2023 John Wiley & Sons Ltd. Published 2023 by John Wiley & Sons Ltd.

27 Observation of Auer rods in crushed cells in acute promyelocytic leukemia

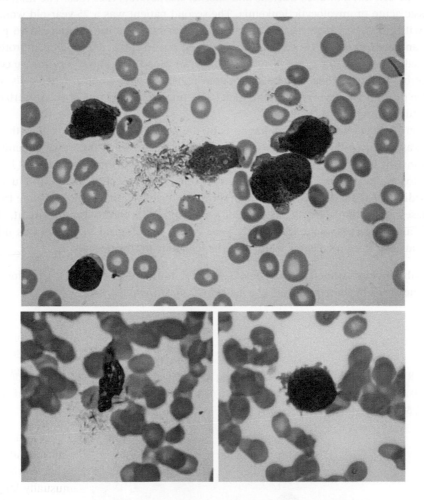

The morphological recognition of acute promyelocytic leukemia is of critical importance because of the need for early specific treatment. However, cytological features are quite variable and in patients in whom the cells have scanty hyperbasophilic cytoplasm, or in whom granules are so densely packed that it is difficult to distinguish cytoplasm from nucleus, recognition of the key features can be difficult. We should like to highlight the value of examination of crushed leukemic cells in blood and bone marrow films from patients with acute promyelocytic leukemia. Two patients are presented here, both with confirmed *PML::RARA* fusion in whom the detection of multiple Auer rods surrounding the nuclei of crushed cells facilitated a morphological diagnosis that was otherwise difficult.

The first patient was a 40-year-old woman who presented with bruising and nose bleeds. Her blood count showed Hb 87 g/l, WBC 8.2×10^9/l, neutrophils 0.9×10^9/l, primitive myeloid cells 6.4×10^9/l and platelet count 10×10^9/l. A coagulation screen including fibrinogen concentration was normal. Both the blood film and bone marrow aspirate showed some crushed cells with

Hematology: 101 Morphology Updates, First Edition. Barbara J. Bain.
© 2023 John Wiley & Sons Ltd. Published 2023 by John Wiley & Sons Ltd.

multiple Auer rods surrounding the nuclei. The bone marrow image (top image) shows four promyelocytes, one crushed with Auer rods and the others intact with hyperbasophilic blebbed cytoplasm; one cell has a bilobed nucleus and another has several Auer rods. The final diagnosis was of classic acute promyelocytic leukemia. The second patient was a 25-year-old woman who presented with tiredness, menorrhagia and bruising. Her blood count showed Hb 96 g/l, WBC 2.9×10^9/l and platelet count 16×10^9/l. A coagulation screen showed a prolonged prothrombin time, reduced fibrinogen (0.6 g/l) and raised D-dimer. A blood film showed primitive cells with scanty hyperbasophilic, blebbed cytoplasm (bottom right) but with some crushed cells revealing the presence of multiple Auer rods (bottom left). The final diagnosis was of the hyperbasophilic variant of acute promyelocytic leukemia.[1]

Facilitation of a cytomorphological diagnosis by the detection of crushed cells with multiple Auer rods was of particular importance in indicating the diagnosis in these two patients for different reasons. In the first patient, there was no coagulation abnormality and no t(15;17) was present, cytogenetic analysis showing only 46,XX,del(7)(q22q36)[7]46,XX[13]. Fluorescence *in situ* hybridization (FISH) analysis, however, showed a fusion signal on chromosome 15 in 17 of 20 metaphases, indicating a cryptic insertion of *RARA* into *PML*. In the second patient, although cytogenetic and molecular genetic features were typical, cytological diagnosis was otherwise difficult because of the cytoplasmic features of the leukemic promyelocytes.

Original publication: Ahmed SO, Deplano S, May PC and Bain BJ (2013) Observation of Auer rods in crushed cells in acute promyelocytic leukemia. *Am J Hematol*, **88**, 236.

Reference

1 McKenna RW, Parkin J, Bloomfield CD, Sundberg RD and Brunning RD (1982) Acute promyelocytic leukaemia: a study of 39 cases with identification of a hyperbasophilic microgranular variant. *Br J Haematol*, **50**, 201–214.

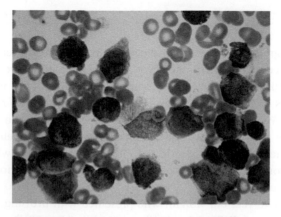

Further image of the bone marrow of the first patient showing multiple delicate Auer rods in a crushed cell.

28 Alpha chain inclusions in peripheral blood erythroblasts and erythrocytes

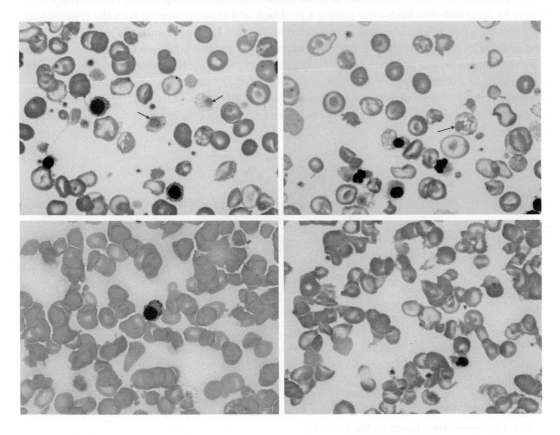

Because of the imbalance between the rate of synthesis of α and β globin chains, α chains precipitate within erythroblasts in the bone marrow in β thalassemia intermedia and major. Sometimes they are also apparent in the peripheral blood in both erythroblasts and erythrocytes. They are particularly likely to be seen in erythrocytes if the pitting function of the spleen has been lost as a result of splenectomy. The top images show the peripheral blood film of a young woman with β thalassemia major who has had a splenectomy and who, because of psychological problems, has been receiving suboptimal transfusion therapy. Her blood count showed a total nucleated cell count of 185×10^9/l, WBC 7.4×10^9/l, Hb 63 g/l, MCV 68 fl, MCH 20 pg and MCHC 291 g/l. The blood film is dimorphic (post-transfusion) and in addition shows nucleated red blood cells (two with lobulated nuclei), Howell–Jolly bodies, Pappenheimer bodies and an acanthocyte. There are three erythrocytes containing α chain inclusions (arrows).

The lower images are from a patient with β thalassemia intermedia (also known as non-transfusion-dependent β thalassemia) with an unusual molecular mechanism. This 10-year-old African patient had compound heterozygosity for $β^0$ thalassemia and hereditary persistence of fetal hemoglobin (HPFH2). The blood count showed Hb 94 g/l, MCV 62 fl, MCH 18.5 pg and MCHC 297 g/l. The blood film (lower images above) shows anisocytosis, poikilocytosis, hypochromic cells and Pappenheimer bodies. There are two erythroblasts, both of which contain an α chain

Hematology: 101 Morphology Updates, First Edition. Barbara J. Bain.
© 2023 John Wiley & Sons Ltd. Published 2023 by John Wiley & Sons Ltd.

inclusion. In the left image the cytoplasm surrounding the inclusion is packed with Pappenheimer bodies.

The presence of α chain inclusions in these thalassemic conditions provides visual evidence of severe chain imbalance, both cases having a total lack of β chain synthesis, with the condition in the second patient being only partly ameliorated by increased γ chain synthesis.

Occasionally, β thalassemia major is first diagnosed unexpectedly on bone marrow examination when α chain inclusions are discovered. The image below is from the bone marrow of a 3-month-old baby girl from a rural area of Iraq in whom a bone marrow examination was done because leishmaniasis was suspected.[1] Erythropoiesis was micronormoblastic, hyperplastic and dysplastic with inadequately hemoglobinized cytoplasm; the image below shows three erythroblasts with α chain inclusions (arrows).

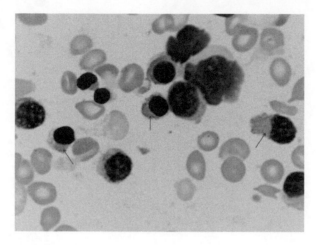

Original publication: Bain BJ (2021) Alpha chain inclusions in peripheral blood erythroblasts and erythrocytes. *Am J Hematol*, **96**, 630–631.

Reference

1 Abdulsalam AH, Sabeeh N and Bain BJ (2011) Diagnosis of beta thalassemia major from bone marrow morphology. *Am J Hematol*, **86**, 187.

29   Dyserythropoiesis in visceral leishmaniasis

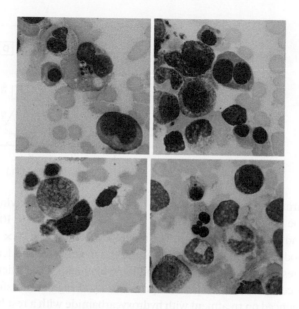

In making a diagnosis of myelodysplastic syndrome it is important to be aware of the numerous causes of myelodysplasia other than a hematopoietic neoplasm. Visceral leishmaniasis (kala azar) is one of the many causes of dyserythropoiesis.[1] This images show a histiocyte containing leishmania (top left), a binucleate erythroblast (top right), a trinucleate erythroblast (bottom left) and abnormal erythroblasts showing, respectively, a detached nuclear fragment and four nuclei in an erythroblast that also shows basophilic stippling (bottom right). Other abnormalities that were present included cytoplasmic bridging and binucleated cells with nuclei of unequal size. The reticulocyte count may be inappropriately low in patients with leishmaniasis with erythroid hyperplasia, indicating that erythropoiesis is ineffective as well as morphologically dysplastic. Since parasites may be infrequent it is important to carry out a diligent search when confronted with cytopenia and dyserythropoiesis in a person who has resided in or visited a country where this parasite is endemic. Observation of the 'double dot' appearance of the nucleus and kinetoplast (top left) helps in identification of *Leishmania donovani*. Organisms may be identified extracellularly, as a result of cell rupture, as well as within macrophages.

The bone marrow in leishmaniasis may show other abnormal features including increased plasma cells, a diffuse increase in histiocytes, hemophagocytosis and granuloma formation.

Original publication: Bain BJ (2010) Dyserythropoiesis in visceral leishmaniasis. *Am J Hematol*, **85**, 781.

Reference

1 Kopterides P, Halikias S and Tsavaris N (2003) Visceral leishmaniasis masquerading as myelodysplasia. *Am J Hematol*, **74**, 198–199.

30 Compound heterozygosity for hemoglobins S and D

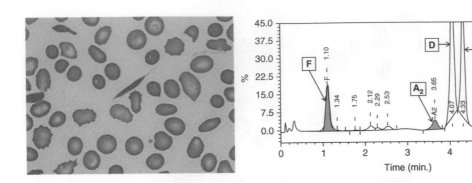

A 15-year-old refugee from Iraq was referred to our hospital shortly after arriving in the UK. He had a long history of fatigue and bone pain. His blood count showed WBC 6×10^9/l, Hb 92 g/l, MCV 99.6 fl, MCH 35.1 pg and platelets 226×10^9/l. The reticulocyte count was 80×10^9/l. His blood film (left image) showed moderate numbers of sickle cells, some macrocytes and features of hyposplenism (Howell–Jolly bodies, Pappenheimer bodies and large platelets). The patient was unaware that he suffered from sickle cell disease but reported a prior history of regular transfusion. Before leaving Iraq he had been commenced on treatment with hydroxycarbamide with a resultant improvement in his transfusion requirement. He had also been taking deferasirox and his serum ferritin was 706 µg/l.

High performance liquid chromatography (right, BioRad Variant II), cellulose acetate electrophoresis at alkaline pH and acid agarose electrophoresis showed hemoglobins S and D. Hemoglobin S was 35.1%, hemoglobin D 42.3%, hemoglobin F 14.3% and hemoglobin A_2 4.1%. Beta globin gene analysis by Sanger sequencing confirmed compound heterozygous mutations of *HBB* c.20A>T, p.Glu7Val (historically described as beta codon 6 GAG>GTG Glu>Val) encoding hemoglobin S and c.364G>C, p.Glu122Gln (historically described as beta codon 121 GAA>CAA Glu>Gln) encoding hemoglobin D Punjab/D Los Angeles.

It is important to remember that 'sickle cell disease' is not synonymous with sickle cell anemia but encompasses also a number of symptomatic compound heterozygous states. S/D disease was first recognized in 1951 in a 'white American' family from Los Angeles with the newly identified variant hemoglobin being named 'd'.[1] Hemoglobin D Punjab is unusual in Middle Eastern populations where hemoglobin D Iran is more often found. It is most prevalent (~2%) in the Sikh population of northwest India but also occurs in Gujarati (~1%), Pakistani, Thai, African American, Afro-Caribbean, northern European Caucasian, Yugoslavian, Turkish and Chinese populations. Because of its possible interaction with hemoglobin S, hemoglobin D Punjab should be identified in antenatal screening programs.

Original publication: Lund K, Chakravorty S, Toma S and Bain BJ (2015) Compound heterozygosity for hemoglobins S and D. *Am J Hematol*, **90**, 842.

Reference

1 Itano HA (1951) A third abnormal hemoglobin associated with hereditary hemolytic anemia. *ProcNatl Acad Sci USA*, **37**, 775–784.

31   Granular B lymphoblastic leukemia

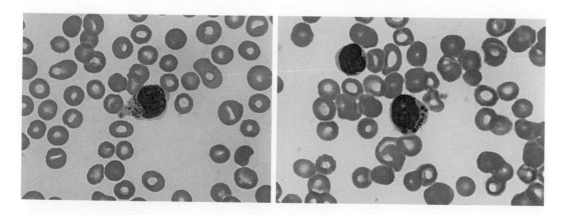

A 13-year-old Indian boy presented with extreme lethargy and weakness. On examination there was no hepatosplenomegaly or lymphadenopathy. His blood count showed WBC 18.8×10^9/l, Hb 107 g/l and platelet count 70×10^9/l. Examination of a blood film led to a suspicion of granular lymphoblastic leukemia. There were 60% blast cells with a moderately high nucleocytoplasmic ratio, delicate chromatin and 1–3 nucleoli. Many of the blast cells contained prominent azurophilic cytoplasmic granules (images). There were no Auer rods and granulocyte morphology was normal. The suspected lineage was confirmed by flow cytometric immunophenotyping which showed the blast cells to express CD45 (weak), CD34, CD10, CD19, CD20, cytoplasmic (c) CD79a, CD33 and HLA-DR. There was no expression of myeloperoxidase, cCD3, CD7, CD13, CD64 or CD117. Cytogenetic analysis showed a normal male karyotype. Reverse transcriptase for *BCR::ABL1* did not detect p210, p190 or p230 transcripts.

Granular acute lymphoblastic leukemia (ALL) is uncommon. Making a distinction from acute myeloid and mixed phenotype acute leukemia is important and requires myeloperoxidase cytochemistry or immunophenotyping. Most cases are pro-B or common ALL but T-lineage cases have also been reported. In addition to B-lineage-specific markers, there can be aberrant expression of CD13[1] or CD33.[2,3] There is no specific cytogenetic association and the prognostic significance, if any, is uncertain.

Original publication: Hazarika B and Bain BJ (2023) Granular B lymphoblastic leukemia. *Am J Hematol*, **98**, 210–211.

References

1 Tembhare PR, Subramanian PG, Sehgal K, Yajamanam B, Kumar A and Gujral S (2009) Hypergranular precursor B-cell acute lymphoblastic leukemia in a 16-year-old boy. *Indian J Pathol Microbiol*, **52**, 421–423.
2 Zhang J, Li M and He Y (2019) Granular B-lineage acute lymphoblastic leukaemia mimicking acute myeloid leukaemia. *Br J Haematol*, **184**, 894.
3 Loyola I, Cadahía P and Trastoy A (2018) Hypergranular lymphoblastic leukaemia. *Br J Haematol*, **182**, 466.

Hematology: 101 Morphology Updates, First Edition. Barbara J. Bain.
© 2023 John Wiley & Sons Ltd. Published 2023 by John Wiley & Sons Ltd.

32 Hyposplenism in adult T-cell leukemia/lymphoma

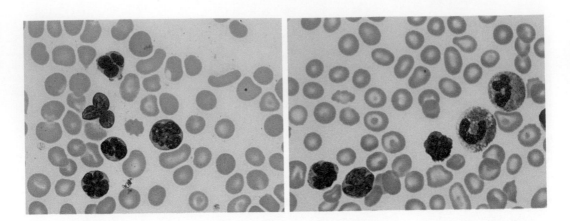

A 76-year-old man of Trinidadian origin presented with a 6-month history of worsening shortness of breath on exertion and a generalized itchy maculopapular rash over the previous 6 weeks. There were no B symptoms or other abnormal physical findings. His general practitioner requested a blood count which showed WBC 24.3×10^9/l, Hb 153 g/l, platelet count 216×10^9/l and lymphocytes 15×10^9/l; his corrected calcium was 2.44 mmol/l and lactate dehydrogenase 430 iu/l. The blood film revealed pleomorphic lymphocytes, some with polylobated nuclei, suggesting a diagnosis of adult T-cell leukemia/lymphoma (ATLL) with leucoerythroblastic changes, together with target cells and Howell–Jolly bodies indicative of hyposplenism (images).

Flow cytometry showed an excess of lymphocytes, of which 85% expressed weak CD3 together with CD2, CD4, CD5 and CD25, with CD7 and CD8 being negative. The immunophenotype was in keeping with ATLL and serology was positive for human T-cell lymphotropic virus 1 (HTLV-1). The proviral load was measured at 60% (proportion of infected mononuclear cells). Positron emission tomography/computed tomography showed small, metabolically active lymph nodes in both axillae, measuring up to 11 mm, with a standardized uptake value (SUV) maximum of 3.2, and bilateral inguinal nodes measuring up to 9 mm, with an SUV maximum of 4.2. Although the spleen was not enlarged it had an SUV maximum of 3.9, which was higher than the liver background of 3.1, suggestive of splenic involvement.

Hyposplenism is usually the result of splenectomy, splenic atrophy (as in celiac disease) or recurrent splenic infarction followed by atrophy and fibrosis (as in sickle cell disease). Functional hyposplenism can also result from splenic overload in acute hemolytic anemia. It is unusual to see hyposplenic features, such as Howell–Jolly bodies and target cells, in the presence of a normal-sized spleen. We postulate that in this patient it resulted from lymphomatous infiltration with replacement of normal splenic tissue. With subsequent disease progression, the patient's spleen became palpable.

Original publication: Cook L, Al-Yousuf H, Mohamedbhai S and Bain BJ (2022) Hyposplenism in adult T-cell leukemia/lymphoma. *Am J Hematol*, **97**, 966–967.

Hematology: 101 Morphology Updates, First Edition. Barbara J. Bain.
© 2023 John Wiley & Sons Ltd. Published 2023 by John Wiley & Sons Ltd.

33 Voxelotor in sickle cell disease

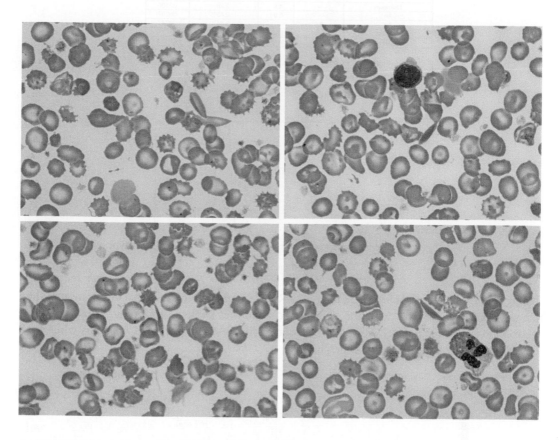

A 45-year-old woman with sickle cell/ß thalassemia due to the *HBB* c.92+5G>C severe ß$^+$ thalassemia mutation, who had been on a clinical trial of voxelotor for 4 years, was assessed clinically and hematologically during follow-up. Concomitantly she was taking hydroxycarbamide (hydroxyurea) at a stable dosage. During this period of combined therapy her Hb had risen from 86 to 103 g/l and MCHC from 327 to 346 g/l with a varying MCV of 79–90 fl. Her blood film, having been typical of sickle cell disease pre-voxelotor, now showed only rare sickle cells and small numbers of boat-shaped or spindle-shaped cells, some of which appeared dense (images above). In addition there were hyposplenic features (Howell–Jolly bodies, target cells, Pappenheimer bodies, acanthocytes and large platelets) together with nucleated red blood cells and macrocytes, some of which were polychromatic.

Peak Name	Calibrated Area %	Area %	Retention Time (min)	Peak Area
F	16.9*	– – –	1.07	87071
P2	– – –	0.5	1.27	3274
Unknown	– – –	4.9	2.28	32100
Ao	– – –	5.8	2.45	38207
A2	6.0*	– – –	3.64	38747
Unknown	– – –	13.3	4.20	87534
S-window	– – –	56.4	4.39	371621

Total Area: 658, 555*

F Concentration = 16.9*%
A2 Concentration = 6.0*%

*Values outside of expected ranges
Analysis comments:

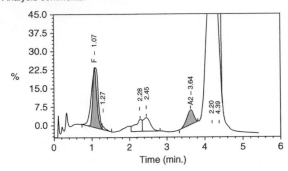

High performance liquid chromatography (BioRad Variant II, above) showed duplication of the hemoglobin S and F peaks with the normal and abnormal peaks overlapping. Capillary zone electrophoresis (Sebia, below) showed duplication of the hemoglobin F and A_2 bands while the hemoglobin S peak was of abnormal shape.

Sample # : **85**   Date : **05/11/2021**   ID : **4347104505**

Depart. :   Birth. :

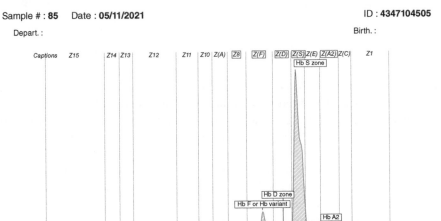

Haemoglobin Electrophoresis

Name	%	Normal Values %
Z8 zone	1.3	
Hb F or Hb variant	12.8	
Hb D zone	2.7	
Hb S zone	78.3	
Hb A2 zone	2.3	
Hb A2	2.6	

Voxelotor is a first-in-class allosteric modifier of hemoglobin that increases the oxygen affinity of hemoglobin S with inhibition of sickling and improvement in Hb and markers of hemolysis. Whether the anti-sickling benefits of voxelotor might be offset by reduced oxygen delivery to tissues remains controversial.[1] The drug forms complexes with the N terminus of the α globin chain. Sickling is inhibited to the extent that a diligent search of a blood film may be needed before any sickle cells are identified. Voxelotor alters the structure not only of hemoglobin S but also of hemoglobins A, F and A_2. The altered characteristics can be detected on isoelectric focusing as well as by HPLC and capillary electrophoresis.[2] The altered hemoglobin sometimes overlaps with the corresponding unaltered hemoglobin and sometimes forms a distinct peak or band. There may be a difference between high performance liquid chromatography (HPLC) and capillary electrophoresis as to whether or not peaks are distinct. We therefore recommend that when there are two peaks, the laboratory uses the format 'hemoglobin S x%, altered hemoglobin S y%', with similar wording when hemoglobin A is present and there are two distinct hemoglobin A peaks. When the peaks overlap and have been quantified together, the wording can be 'hemoglobin S plus altered hemoglobin S, x%'.

Original publication: Bain BJ, Myburgh J, Hann A and Layton AM (2022) Voxelotor in sickle cell disease. *Am J Hematol*, **97**, 830–832.

References

1   Henry EER, Metaferia B, Li Q, Harper J, Best RB, Glass KE *et al.* (2021) Treatment of sickle cell disease by increasing oxygen affinity of hemoglobin. *Blood*, **138**, 1172–1181.
2   Rutherford NJ, Thoren KL, Shajani-Yia Z and Colbya JM (2018) Voxelotor (GBT440) produces interference in measurements of hemoglobin S. *Clin Chim Acta*, **482**, 57–59.

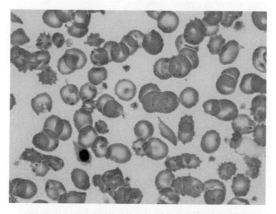

A further image showing a single boat-shaped cell, a nucleated red blood cell, echinocytes, a target cell and Papperheimer bodies.

34 The importance of a negative image

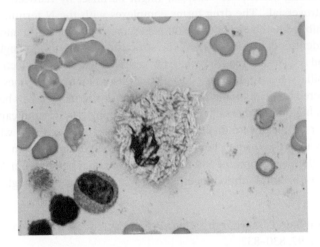

An HIV-positive woman had a bone marrow aspirate performed for investigation of fever. This showed the negative images of bacilli, both within macrophages and free. In this clinical context, the observation of macrophages stuffed with bacilli on a Romanowsky-stained film is diagnostic of atypical mycobacterial infection and means that treatment can be started while awaiting culture results. An acid-fast stain is also positive. Such numerous bacilli are not a feature of infection by *Mycobacterium tuberculosis*. Negative images of mycobacteria have also been reported in leprosy.

Original publication: Bain BJ (2008) The importance of a negative image. *Am J Hematol*, **83**, 410.

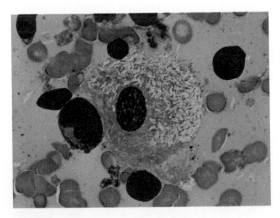

A further macrophage in the patient's bone marrow showing the negative images of mycobacteria.

Hematology: 101 Morphology Updates, First Edition. Barbara J. Bain.
© 2023 John Wiley & Sons Ltd. Published 2023 by John Wiley & Sons Ltd.

35 Seeing what isn't there

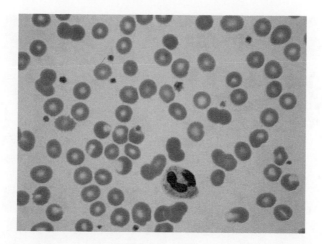

This image shows the blood film of a patient who was hospitalized with renal failure. At first glance there appeared to be a number of unusually shaped red cells. Initially a hemolytic anemia was suspected and investigation was commenced, but on further reflection it was realized that the apparent holes in the red cells were likely to represent overlying cryoglobulin. This possibility was confirmed and was found to be due to type II cryoglobulinemia in a patient with hepatitis C infection. The hepatitis C infection was, in turn, the cause of the renal failure. Cryoglobulin can be weakly basophilic, in which case it is easily recognized on a blood film. However, it is also necessary to recognize the negative image of unstained cryoglobulin, which can be an important clue to the diagnosis of hepatitis C infection.

Original publication: Bain BJ and Patel B (2008) Seeing what isn't there. *Am J Hematol*, **83**, 504.

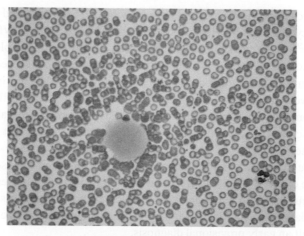

A low power view of the blood film of another patient with cryoglobulinemia showing a large extracellular deposit of cryoglobulin.

Hematology: 101 Morphology Updates, First Edition. Barbara J. Bain.
© 2023 John Wiley & Sons Ltd. Published 2023 by John Wiley & Sons Ltd.

36 A young woman with sudden onset of a severe coagulation abnormality

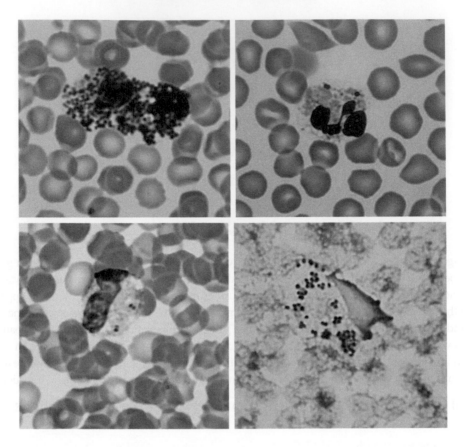

Blood samples were received from a 29-year-old woman who had been admitted to the resuscitation room of an emergency department. Her blood count showed Hb 152 g/l, WBC 9.6×10^9/l and platelet count 39×10^9/l. A citrate sample for coagulation tests was unclottable. An urgent blood film (images) confirmed thrombocytopenia and showed heavily vacuolated neutrophils and a left shift. Many of the neutrophils contained organisms (top left and right) and in some cells it was apparent that they were diplococci (bottom left). A Gram stain showed the organism to be Gram-negative (bottom right) and a provisional diagnosis of meningococcal septicaemia with disseminated intravascular coagulation was made. A latex agglutination test for meningococcal antigen confirmed the diagnosis and the organism was typed as *Neisseria meningitidis* group C. Despite appropriate early and vigorous management the patient required several amputations and succumbed to the effects of the meningococcal infection 3 weeks after presentation.

Peripheral blood neutrophils in meningococcal septicemia show characteristic features of toxic granulation, degranulation and marked vacuolation. Organisms are often detectable within the neutrophils and permit an early provisional diagnosis.

Original publication: Uprichard J and Bain BJ (2008) A young woman with sudden onset of a severe coagulation abnormality. *Am J Hematol*, **83**, 672.

Hematology: 101 Morphology Updates, First Edition. Barbara J. Bain.
© 2023 John Wiley & Sons Ltd. Published 2023 by John Wiley & Sons Ltd.

37 Immature *Plasmodium falciparum* gametocytes in bone marrow

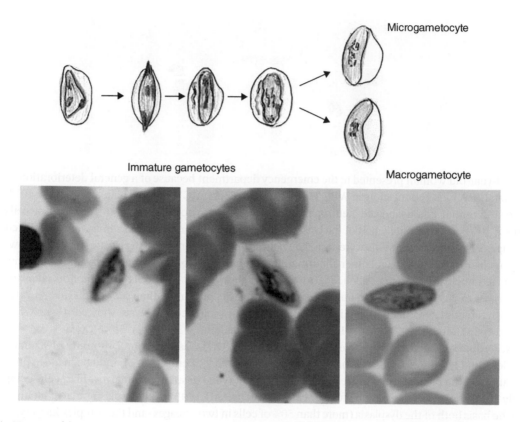

A 50-year-old Iraqi man presented with splenomegaly and pyrexia of unknown origin. A bone marrow aspirate was done as part of the investigations and unexpectedly showed *Plasmodium falciparum* parasites. Malaria had not been suspected as this condition is now rare in Iraq but it subsequently transpired that the patient had recently visited Pakistan.

The gametocytes that are observed in the bone marrow differ from those that are observed in the blood, being less mature.[1] The immature gametocytes that are seen include some that are sail-shaped, spindle-shaped or oval (top image) rather than the crescent-shaped macrogametocyte and sausage-shaped microgametocyte that are usually observed in the blood. This reflects the fact that gameto-cytes develop in the internal organs, including the bone marrow, rather than in the circulating blood. The bone marrow of this patient (bottom images) showed sail-shaped (left), spindle-shaped (center) and oval (right) immature gametocytes. Some mature gametocytes were also present.

Original publication: Abdulsalam AH, Sabeeh N and Bain BJ (2010) Immature *Plasmodium falciparum* gametocytes in bone marrow. *Am J Hematol*, **85**, 943.

Reference

1 Smalley ME, Abdalla S and Brown J (1980) The distribution of *Plasmodium falciparum* in the peripheral blood and bone marrow in Gambian children. *Trans Roy Soc Trop Med Hyg*, **75**, 103–105.

38   Acute myeloid leukemia with myelodysplasia-related changes showing basophilic differentiation

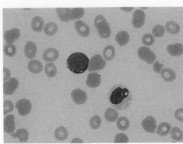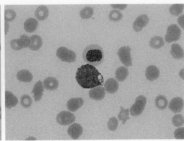

An 84-year-old woman presented to the emergency department because of a general deterioration in her health. A blood count showed WBC $23.4 \times 10^9/l$, Hb 52 g/l, MCV 89 fl and platelet count $9 \times 10^9/l$. The differential count showed neutrophils $1.6 \times 10^9/l$, lymphocytes $1.3 \times 10^9/l$ and blast cells $20.9 \times 10^9/l$. Approximately half of the blast cells had basophilic granules and some were vacuolated (images). No mature basophils were seen. The neutrophils were almost all dysplastic (images), showing nuclear hypolobation and other abnormalities of nuclear shape, hypogranularity and occasional detached nuclear fragments. A bone marrow aspirate showed 12% blast cells and 2 months later this had risen to 22% blast cells. Granulopoiesis was dysplastic and more than 50% of megakaryocytes were hypolobulated, including micromegakaryocytes. Cytogenetic analysis showed 41-42,XX,-3, add(5)(p13),add(6)(q?25),-7,-10,+11,-15,-17[cp10]/44-45,idem,+3,-4,+mar[5]/46XX[5].

According to the 2016 World Health Organization (WHO) classification of tumours of haematopoietic and lymphoid tissues,[1] a diagnosis of acute myeloid leukemia (AML) could be made at presentation on the basis of the peripheral blood blast count (90%) despite bone marrow blast cells being only 12%. The leukemia could be further categorized as AML with myelodysplasia-related changes on the basis both of the dysplasia (more than 50% of cells in two lineages) and the complex karyotype. Despite the prominent basophilic differentiation, the diagnosis is not acute basophilic leukemia since, in the hierarchical WHO classification, the myelodysplasia-related changes take precedence.

Basophilic differentiation can be clinically relevant since occasional patients have features of histamine excess – urticaria and peptic ulceration[2] – and an anaphylactoid reaction following chemotherapy has been reported.[3]

Original publication: Wells R, Williams B and Bain BJ (2014) Acute myeloid leukemia with myelodysplasia-related changes showing basophilic differentiation. *Am J Hematol*, **89**, 1082.

References

1 Arber DA, Brunning RD, Orazi A, Bain BJ, Porwit A, Le Beau MM and Greenberg PL (2017) Acute myeloid leukaemia with myelodysplasia-related changes. In: Swerdlow SH, Campo E, Harris NL, Jaffe ES, Pileri SA, Stein H and Thiele J (eds). *WHO Classification of Tumours of Haematopoietic and Lymphoid Tissues*. IARC Press, Lyon, pp. 150–152.

2 Duchayne E, Demur C, Rubie H, Robert A and Dastugue N (1999) Diagnosis of acute basophilic leukemia. *Leuk Lymphoma*, **32**, 269–278.

3 Bernini JC, Timmons CF and Sandler ES (1995) Acute basophilic leukemia in a child: anaphylactoid reaction and coagulopathy secondary to vincristine-mediated degranulation. *Cancer*, **75**, 110–114.

39   Thiamine-responsive megaloblastic anemia in an Iraqi girl

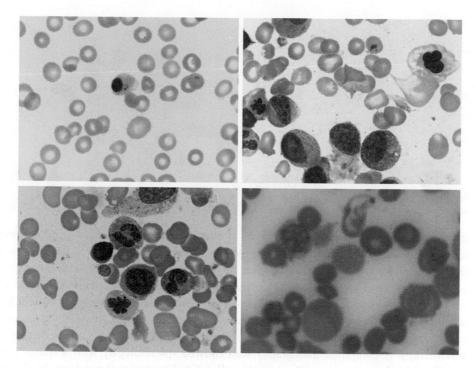

A 27-month-old female presented with insulin-dependent diabetes mellitus, chronic anemia, deafness and cardiomyopathy. A full blood count showed WBC 4×10^9/l, Hb 71 g/l and platelet count 195×10^9/l. A differential count showed neutrophils 39%, lymphocytes 48%, monocytes 7%, eosinophils 5%, myelocytes 1% and nucleated red blood cells 2%. A blood film (top left image) showed marked anisocytosis with some very large macrocytes, mild poikilocytosis and the presence of circulating megaloblasts. A bone marrow aspirate was hypercellular (absent fat cells) with dysplastic megaloblastic erythropoiesis, erythroid vacuolation and punctate basophilia (top right and bottom left). An iron stain showed markedly increased iron stores with many ring sideroblasts (bottom right). Granulopoiesis and megakaryopoiesis were normal.

The child was the daughter of consanguineous parents. An older brother was healthy but two older sisters had died of diabetic ketoacidosis and a 16-month-old sister had similar clinical and hematological features to those seen in the patient.

On the basis of the clinical history and the bone marrow findings, a diagnosis of thiamine-responsive megaloblastic anemia was made. This autosomal recessive syndrome is characterized by sensorineural deafness, diabetes mellitus and thiamine-responsive megaloblastic anemia with ring sideroblasts. Genetic analysis was not done in this family but the condition is known to result from mutation in the *SLC19A2* gene. A significant proportion of the families described until now have been of Middle Eastern origin. The patient and her surviving sister showed a hematological response to thiamine therapy and their insulin requirements reduced.

Original publication: Abdulsalam AH, Sabeeh N, Ibrahim ZI and Bain BJ (2014) Thiamine-responsive megaloblastic anemia in an Iraqi girl. *Am J Hematol*, **89**, 659.

Hematology: 101 Morphology Updates, First Edition. Barbara J. Bain.
© 2023 John Wiley & Sons Ltd. Published 2023 by John Wiley & Sons Ltd.

40 Teardrop poikilocytes in metastatic carcinoma of the breast

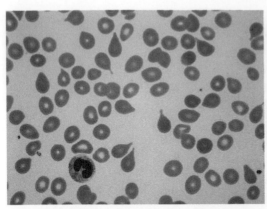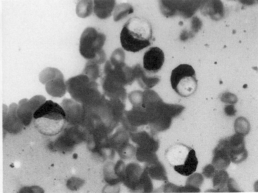

A 77-year-old woman presented with symptoms of anemia. No splenomegaly or other specific abnormality was detected on examination. Her blood count showed WBC 5.3×10^9/l, Hb 86 g/l, MCV 92 fl and platelet count 105×10^9/l. Liver function tests and serum calcium, phosphate, alkaline phosphatase and creatinine were normal. A blood film was leucoerythroblastic with numerous teardrop poikilocytes (left image). There were no dysplastic features. The patient had a previous history of adenocarcinoma of the left breast 22 years previously, treated by mastectomy and chemotherapy, and anorectal carcinoma 6 years previously, treated with chemotherapy and pelvic radiotherapy. A computed tomography scan showed splenomegaly with a spleen length measuring 14 cm, and numerous small, relatively well-defined lucencies throughout the skeleton. A bone marrow aspirate yielded only blood but an imprint from a trephine biopsy specimen showed adenocarcinoma cells containing large globules of mucin (right), indicating a diagnosis of metastatic carcinoma of the breast. Serum Ca 15-3 (carcinoma antigen 15.3) was 669 kunits/ml (normal <30). Trephine biopsy sections showed an infiltrate composed of round/oval cells, which had signet ring morphology and were present in clusters. These cells stained positively for CAM 5.2, MNF116, estrogen receptor, CK7 and GCDFP, and negatively for CK20, confirming a diagnosis of metastatic breast carcinoma. There was also marrow fibrosis and reduced myelopoiesis.

The blood film in carcinoma metastatic to the bone marrow can simulate that of primary myelofibrosis although the film of primary myelofibrosis may show additional abnormalities such as giant or hypogranular platelets, dysplastic features in leucocytes and occasional circulating micromegakaryocytes and megakaryocyte nuclei. Even a remote history of carcinoma of the breast is relevant in a patient with a leucoerythroblastic blood film since metastases may appear many years after the primary tumor.

Original publication: Bain BJ, Farah N and Flora R (2014) Teardrop poikilocytes in metastatic carcinoma of the breast. *Am J Hematol*, **89**, 557.

41 A blood film that could have averted a splenectomy

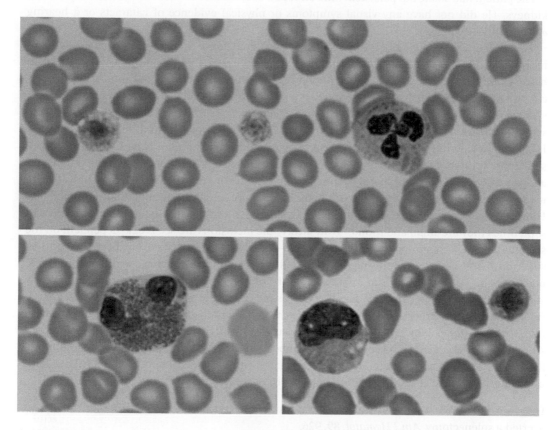

A 31-year-old man who attended the emergency department for a relatively minor ailment was found to have an elevated serum creatinine (142 µmol/l) and thrombocytopenia (platelet count 34×10^9/l). On follow-up testing, the platelet count, as measured by an automated counter, was 16×10^9/l. The blood film showed very large platelets with the platelet count appearing much higher than the instrument count. Myeloid cells – neutrophils, eosinophils and monocytes – showed inclusions that resembled Döhle bodies but were larger and more angular, pointing to a MYH9-related disease (images). Few basophils were visualized on the blood film.

The patient had been generally healthy without any bleeding manifestations. His hemostasis had fortunately not been challenged. He had apparently been known to be thrombocytopenic since childhood but had not been offered any diagnosis. His father had also been thrombocytopenic all his life and had undergone splenectomy as a child without any improvement in his platelet count. The propositus was unaware of any definitive diagnosis having been made in his father but believed that immune thrombocytopenia was suspected. Unfortunately the patient's father had died a few years earlier of a brain tumor and hence macrothrombocytopenia could not be confirmed or excluded. We postulate that the father of the propositus suffered from the same condition as his son and had undergone an unnecessary splenectomy subjecting him, from a relatively young age, to the risks of hyposplenism without there being any benefit. A careful examination of the blood film of the father very likely would have averted the splenectomy. Patients with MYH9-related

Hematology: 101 Morphology Updates, First Edition. Barbara J. Bain.
© 2023 John Wiley & Sons Ltd. Published 2023 by John Wiley & Sons Ltd.

disease have not infrequently been subjected to corticosteroids, azathioprine, high dose intravenous immunoglobulin or splenectomy following misdiagnosis.[1]

The patient had stable but persistent mild elevation of serum creatinine with structurally normal kidneys. He did not have any visual symptoms or physical evidence of cataracts or a hearing abnormality.

'MYH9-related disease' is an umbrella term now used for a group of autosomal dominant inherited conditions resulting from mutation in *MYH9*, the gene that encodes non-muscle myosin heavy chain type IIA. Depending on the presence and nature of any associated abnormalities these conditions were previously described as eponymous syndromes – May–Hegglin anomaly, Epstein syndrome, Fechtner syndrome and Sebastian syndrome – with macrothrombocytopenia and sometimes various combinations of nephritis, cataracts and sensorineural deafness. The May–Hegglin or Döhle-like inclusions are present in neutrophils, eosinophils and basophils. It has been stated that they are not present in monocytes[2] but we and others[3] have observed them. Inclusions are generally numerous and sharply defined, often spindle- or crescent-shaped, randomly distributed in the cell rather than near the cell margin and more intensely staining than Döhle bodies. However, the size and shape of the inclusions is dependent on the specific mutation present and sometimes they are small and round.[1] How clearly the inclusions are see is also dependent on the specific type of Romanowsky stain used; a May–Grünwald–Giemsa stain has been stated to be superior to a Wright stain in this regard.[1] At an ultrastructural level, inclusions differ from Döhle bodies, being composed of amorphous deposits of the mutant protein often incompletely surrounded by a single strand of rough endoplasmic reticulum or containing a few clusters of ribosomes and dense rods and spherical particles, which are probably ribosomes. It is important to note that in some patients inclusions are not visualized by light microscopy but can be demonstrated by fluorescence microscopy with an antibody to MYH9.[1,2] The diagnosis of autoimmune thrombocytopenia should therefore always be made with circumspection in patients who have unusually large platelets, even in the absence of visible leucocyte inclusions.

Original reference: Ghara N, Mohamedbhai S and Bain BJ (2014) A blood film that could have averted a splenectomy. *Am J Hematol*, **89**, 926.

References

1 Althaus K and Greinacher A (2009) MYH9-related platelet disorders. *Semin Thromb Hemost*, **35**, 189–203.

2 Saito H and Kunishima S (2011) Historical hematology: May–Hegglin anomaly. *Am J Hematol*, **83**, 304–306.

3 Balduini CL, Pecci A and Savoia A (2011) Recent advances in the understanding and management of MYH9-related inherited thrombocytopenias. *Br J Haematol*, **154**, 161–174.

42 Russell bodies and Mott cells

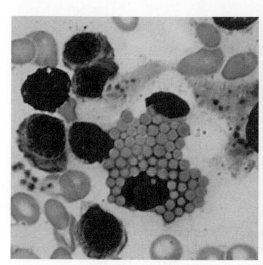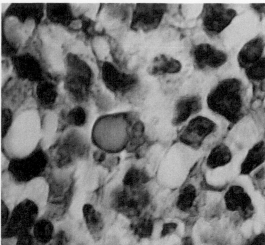

Mott cells are plasma cells that have spherical inclusions packing their cytoplasm. The left image above shows a Mott cell in the bone marrow aspirate of a patient with monoclonal gammopathy of undetermined significance. The term Mott cell derives from a surgeon, F. W. Mott, who described these cells in the brains of monkeys with trypanosomiasis.[1] He himself used the term morular cell (from the Latin *morus*, mulberry) and he recognized that the cells were plasma cells and indicative of chronic inflammation. Although his name has become attached to the cell, Mott was not the first to describe this appearance. The first description was probably that of William Russell in 1890, although he recognized neither the nature of the cell nor the significance of the inclusions. The term 'Russell body' is used inconsistently between countries and even within countries. Some pathologists use it only to refer to a large spherical inclusion displacing the nucleus. Others use the term also to refer to the multiple inclusions within Mott cells. Inspection of the hand-drawn illustrations in Russell's article[2] reveals that he observed multiple spherical inclusions within single cells. Mott cells therefore contain Russell bodies. The right image from the same patient (above) shows a Mott cell containing one large inclusion and a number of smaller ones in a section of a bone marrow trephine biopsy specimen (H&E). The inclusions of Mott cells are now known to represent immunoglobulin within vesicular structures derived from dilated rough endoplasmic reticulum. Mott cells and Russell bodies occur in both reactive plasmacytosis and in plasma cell neoplasms. In plasma cell neoplasms, clonality can be demonstrated by immunohistochemistry (left image below, positive anti-lambda; right image below, negative anti-kappa).[3]

Hematology: 101 Morphology Updates, First Edition. Barbara J. Bain.
© 2023 John Wiley & Sons Ltd. Published 2023 by John Wiley & Sons Ltd.

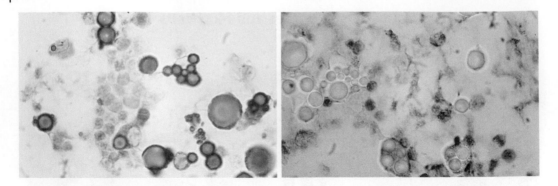

Original publication: Bain BJ (2009) Russell bodies and Mott cells. *Am J Hematol*, **84**, 516.

References

1 Mott FW (1905) Observations on the brains of men and animals infected with various forms of trypanosomes. *Preliminary note. Proc Roy Soc London*, **76**, 235–242.

2 Russell W (1890) An address on a characteristic organism of cancer. *BMJ*, ii, 1358–1360.

3 Bain BJ (2009) Russell bodies. *Am J Hematol*, **84**, 439.

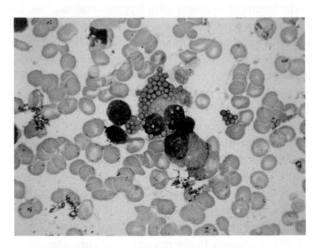

Bone marrow film from another patient showing a binucleated Mott cell.

43   Dutcher bodies

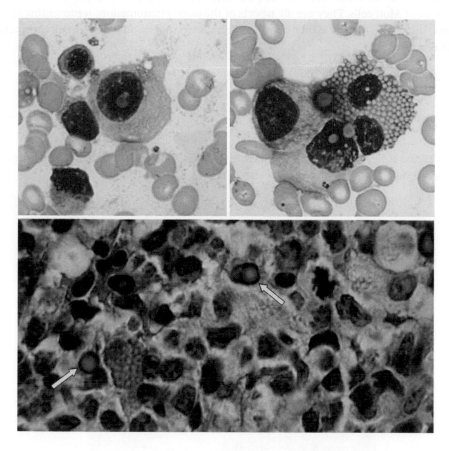

Dutcher bodies were described by Dutcher and Fahey in 1959 as intranuclear inclusions in patients with Waldenström macroglobulinemia.[1] The inclusions are positive on a periodic acid–Schiff (PAS) stain and were present in the cytoplasm as well as the nucleus. They identified the inclusions as glycoprotein and postulated that they might be chemically identical to the macroglobulin in the plasma. They thought that the inclusions developed in the nucleus and were extruded through a ruptured nuclear membrane into the cytoplasm. It is now known that Dutcher bodies are actually cytoplasmic inclusions that are either invaginated into or are overlying the nucleus. The images above from a patient with monoclonal gammopathy of undetermined significance illustrate how cytoplasmic inclusions can appear to be intranuclear. One neoplastic plasma cell (top left) appears to have an intranuclear inclusion (May–Grünwald–Giemsa). However, plasma cells in another cluster (top right) show a variety of appearances; three medium-sized Russell bodies with the same staining characteristics as the inclusion in the cell already observed either overlie the nucleus partly or entirely or are within the cytoplasm. There is also a Mott cell filled with smaller cytoplasmic Russell bodies. A trephine biopsy section from the same patient (bottom) shows two Dutcher bodies that are overlying nuclei and thus appear as intranuclear inclusions (arrows) (H&E).

Hematology: 101 Morphology Updates, First Edition. Barbara J. Bain.
© 2023 John Wiley & Sons Ltd. Published 2023 by John Wiley & Sons Ltd.

Adjacent to the lower Dutcher body-containing plasma cell is a Mott cell with abundant cytoplasmic inclusions obscuring the nucleus.

There are no essential differences between Dutcher bodies, single or multiple Russell bodies and the inclusions of Mott cells. They are all aspects of the same phenomenon, representing spherical cytoplasmic inclusions that are either clearly within the cytoplasm or are overlying the nucleus or invaginated into it.

Original publication: Bain BJ (2009) Dutcher bodies. *Am J Hematol*, **84**, 589.

Reference

1 Dutcher TF and Fahey JL (1959) The histopathology of the macroglobulinemia of Waldenström. *J Natl Cancer Inst*, **22**, 887–917.

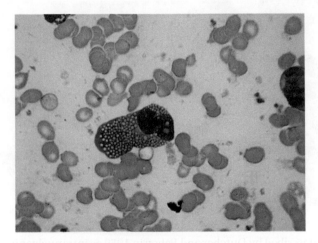

Bone marrow film of another patient showing a binucleated Mott cell. One of the nuclei contains two Dutcher bodies.

44   Acute myeloid leukemia with inv(16)(p13.1q22)

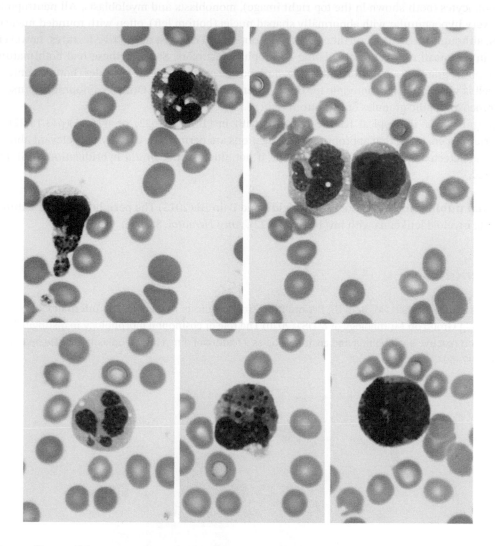

It is usually possible to strongly suspect a diagnosis of acute myeloid leukemia (AML) with inv(16) (p13.1q22) from the characteristic bone marrow morphology. However, it may also be possible to identify distinctive features on careful blood film examination. The findings are those of acute myelomonocytic leukemia with eosinophilia. Although most eosinophils show only minor dysplastic changes, such as cytoplasmic vacuolation and reduction of granules, there may also be some eosinophils and precursors with purple proeosinophilic granules, sometimes large.[1] In addition, neutrophils are dysplastic.

These images are from the blood film of a 48-year-old Somalian woman. Her blood count showed WBC 74.9 × 10⁹/l, Hb 102 g/l and platelet count 46 × 10⁹/l. The cytological abnormalities were typical of this condition. The majority of eosinophils showed mildly dysplastic features, such as the vacuolation seen in the upper cell in the top left image. A minority were more abnormal, being

Hematology: 101 Morphology Updates, First Edition. Barbara J. Bain.
© 2023 John Wiley & Sons Ltd. Published 2023 by John Wiley & Sons Ltd.

very small or showing a lack of nuclear lobulation or the presence of proeosinophilic granules (as seen in the lower cell in the top left image). In addition there were immature monocytes and promonocytes (both shown in the top right image), monoblasts and myeloblasts. All neutrophils were very hypogranular with abnormally shaped nuclei (bottom left), often with rounded nuclear lobes, although classic Pelger–Huët forms were not seen. The most distinctive features, however, were in the small number of circulating eosinophil precursors. Some of these had both mature eosinophilic granules and abnormally large, dark purple proeosinophilic granules (bottom center); this cell is virtually pathognomonic for this cytogenetic abnormality. Other eosinophil precursors had proeosinophilic granules only (bottom right).

Recognition of the typical feature of AML with inv(16)(p13.1q22) or t(16;16)(p13.1;q22) is important in ensuring that appropriate investigations are performed. Usually the relevant abnormality is detected on cytogenetic analysis but, if not, fluorescence *in situ* hybridization (FISH) is required.

Original publication: Sreedhara S, Grinfeld J and Bain BJ (2013) The peripheral blood features of acute myeloid leukemia with inv(16)(p13.1q22). *Am J Hematol*, **88**, 975.

Reference

1 Goasguen JE, Bennett JM, Bain BJ, Brunning R, Zini G, Vallespi M-T *et al.* (The International Working Group on Morphology of MDS) (2020) The role of eosinophil morphology in distinguishing between reactive eosinophilia and eosinophilia as a feature of a myeloid neoplasm. *Br J Haematol*, **191**, 497–504.

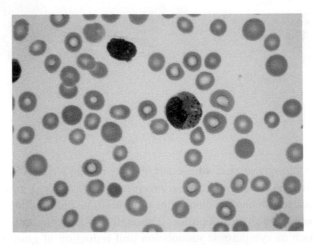

Another image from the same patient showing an eosinophil myelocyte with eosinophilic and proeosinophilic granules.

45   Dysplastic macropolycytes in myelodysplasia-related acute myeloid leukemia

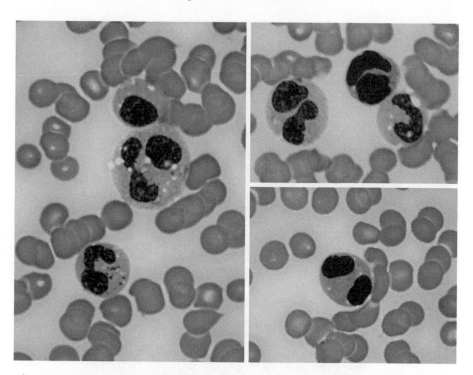

Macropolycytes are not necessarily dysplastic since they are occasionally seen in the blood films of healthy subjects and can appear cytologically normal – apart from being double normal size with a proportionate increase in the size of the nucleus, the degree of nuclear lobulation and, in women, the number of nuclear drumsticks. They have been shown to be tetraploid and thus represent a missed cell division. These cells are increased in number in myeloproliferative neoplasms, during infection and following growth factor administration. They occur also in megaloblastic anemia and HIV infection, when they may show additional cytological abnormalities, including a more open chromatin pattern. Macropolycytes are an uncommon feature of the myelodysplastic syndromes (MDS) and myelodysplasia-related acute myeloid leukemia (AML). The macropolycytes shown here are from a 67-year-old man with myelodysplasia-related AML, who had initially presented 19 months earlier with MDS (specifically, refractory cytopenia with multilineage dysplasia).

The left image shows a macropolycyte flanked by two neutrophils of normal size. All three cells are hypogranular and in addition the upper cell shows an acquired Pelger–Huët anomaly. The nucleus of the macropolycyte is abnormal in shape. The top right image shows two binucleated macropolycytes, one of which has hyperchromatic nuclei, while the bottom right image shows a hypogranular binucleated macropolycyte, with both nuclei showing an acquired Pelger–Huët anomaly.

Dysplastic macropolycytes can occur in reactive conditions and deficiency states but should be recognized as an uncommon feature of myeloid neoplasms, including MDS and AML.

Original publication: Bain BJ (2011) Dysplastic macropolycytes in myelodysplasia-related acute myeloid leukemia. *Am J Hematol*, **86**, 776.

Hematology: 101 Morphology Updates, First Edition. Barbara J. Bain.
© 2023 John Wiley & Sons Ltd. Published 2023 by John Wiley & Sons Ltd.

46   Diagnosis of cystinosis from a bone marrow aspirate

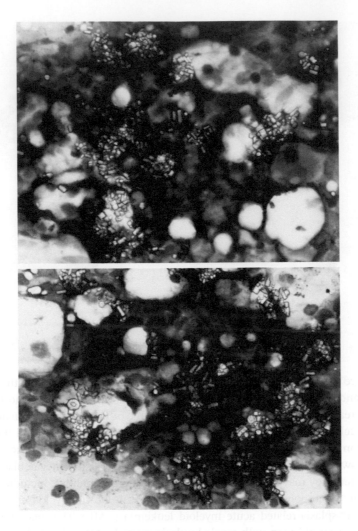

A 4-year-old boy presented with pallor, visual problems and recurrent, multiple, bilateral renal stones that had been unresponsive to (non-specific) medical treatment. He had seen a number of physicians over a period of 2 years without a diagnosis being made. He was found to have pancytopenia, the blood count showing WBC 3.1×10^9/l, Hb 86 g/l and platelet count 102×10^9/l.

A bone marrow aspirate was performed to investigate the pancytopenia and revealed numerous clear cystine crystals, apparently free and within macrophages (images). This rare lysosomal storage disease is readily diagnosed from a bone marrow aspirate because of the presence of distinctive non-staining crystals. The crystals have straight edges and, when seen end on, are hexagonal (bottom image).

Original publication: Abdulsalam AH, Khamis MH and Bain BJ (2013) Diagnosis of cystinosis from a bone marrow aspirate. *Am J Hematol*, **88**, 151.

Hematology: 101 Morphology Updates, First Edition. Barbara J. Bain.
© 2023 John Wiley & Sons Ltd. Published 2023 by John Wiley & Sons Ltd.

47 Emperipolesis in a patient receiving romiplostim

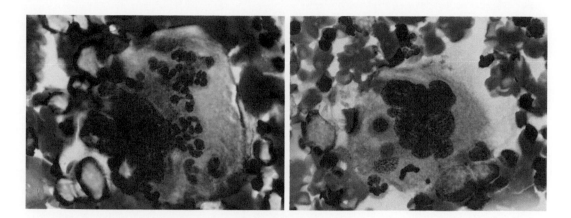

A 22-year-old man with autoimmune thrombocytopenic purpura ('ITP') that was refractory to high dose corticosteroids and high dose intravenous immunoglobulin was prescribed romiplostim, a thrombopoietin mimic, in an increasing dose to a maximum of 10 μg/kg. There was no response of the platelet count, which did not rise above $11 \times 10^9/l$. However, a subsequent bone marrow aspirate showed abundant megakaryocytes with prominent emperipolesis (images). Emperipolesis is the active entry of a living cell into another. In the hemopoietic system it is most often observed as the entry of other cells – granulocytes, erythroblasts, lymphocytes – into the surface-connected canalicular system of megakaryocytes. It differs from phagocytosis in that the cell that penetrates another cell remains intact and is cytologically normal. The term is derived from the Greek e*m* (inside), *peri* (around) and *polemai* (to wander about); it was coined by Humble and colleagues to describe the behaviour of lymphocytes in relation to megakaryocytes, malignant cells and hemopoietic cells in mitosis.[1] Emperipolesis is a common observation if a large number of megakaryocytes are examined and does not seem to have any specific diagnostic significance.[2] It can occur both with neoplastic and with normal megakaryocytes. It is rarely as prominent as in this patient.

Original publication: Cooper N and Bain BJ (2016) Emperipolesis in a patient receiving romiplostim. *Am J Hematol*, **91**, 166.

References

1 Humble JG, Jayne WHW and Pulvertaft RJV (1956) Biological interaction between lymphocytes and other cells. *Br J Haematol*, **2**, 283–294.
2 Rozman C and Vives-Corrons JL (1981) On the alleged diagnostic significance of megakaryocyte 'phagocytosis' (emperipolesis). *Br J Haematol*, **48**, 510.

48 Mechanical hemolysis: a low mean cell volume does not always represent microcytosis

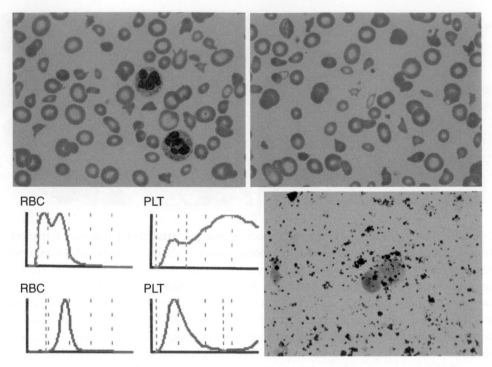

A 56-year-old woman who had had a prosthetic mitral valve replacement 38 years previously presented to cardiology outpatients with a 12-month history of increasing lethargy and shortness of breath. Her Hb and red cell indices on a Sysmex XE-2100 instrument were RBC 3.86×10^{12}/l, Hb 71 g/l, MCV 63.5 fl, MCH 18.4 pg and MCHC 290 g/l. Prior to examination of her blood film, tests for iron deficiency and hemoglobin H disease had been requested. However, her blood film, rather than showing microcytosis, showed very numerous schistocytes including occasional microspherocytes (top images). There were also polychromatic macrocytes and the reticulocyte count was 195×10^9/l (5.1%). No estimate of the red cell distribution width was produced by the automated instrument but inspection of the red cell (RBC) and platelet (PLT) size histograms (bottom left, patient at the top and normal subject at the bottom) showed two erythrocyte populations, one microcytic (representing the schistocytes) and one normocytic; the schistocytes were also apparent on the platelet size histogram. The instrument 'flags' included 'fragments?', 'Dimorphic Population' and 'PLT Abn Distribution'. Haptoglobin was undetectable. Urinary hemosiderin was strongly positive (bottom right, Perls stain). A diagnosis of mechanical hemolytic anemia was made. Transesophageal echocardiography was therefore undertaken and showed a perivalvular leak, necessitating surgery.

It is important to remember that a low MCV may represent red cell fragmentation as well as microcytosis. A low MCV is usual, for example, in hereditary pyropoikilocytosis. It is good practice to examine a blood film and automated instrument plots when the MCV is reduced.

Original publication: Bain BJ, Varu V, Rowley M and Foale R (2015) Mechanical hemolysis: a low mean cell volume does not always represent microcytosis. *Am J Hematol*, **90**, 1179.

Hematology: 101 Morphology Updates, First Edition. Barbara J. Bain.
© 2023 John Wiley & Sons Ltd. Published 2023 by John Wiley & Sons Ltd.

49 Transplant-associated thrombotic microangiopathy

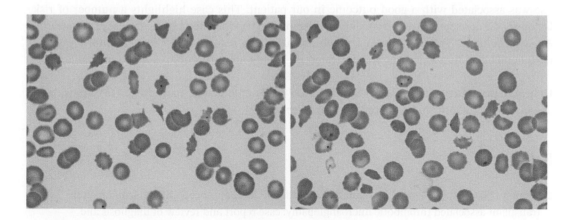

A 16-year-old boy underwent a 9/10 matched unrelated bone marrow transplant for congenital dyserythropoietic anemia. He had previously had a splenectomy. Early post-transplant complications included Epstein–Barr virus and human herpesvirus 6 reactivation. Long-term immune suppression was with ciclosporin. At day 127 post-transplant, he developed weight gain, peripheral edema, ascites and pleural effusions and received three doses of the vascular endothelial growth factor (VEGF) inhibitor bevacizumab for possible capillary leak syndrome. He subsequently developed renal and respiratory failure and seizures with radiological evidence of posterior reversible encephalopathy syndrome. Blood tests showed WBC 0.9×10^9/l, Hb 65 g/l, platelet count 10×10^9/l, bilirubin 54 µmol/l, creatinine 154 µmol/l and lactate dehydrogenase 1175 iu/l; haptoglobin was undetectable. A direct antiglobulin test was negative and the urine protein/creatinine ratio was 213 mg/mmol (normal <20). His blood film showed the features of a microangiopathic hemolytic anemia complicated by hyposplenism and acute kidney injury (images). In addition to thrombocytopenia and red cell fragments, there were nucleated red blood cells, Howell–Jolly bodies, Pappenheimer bodies, echinocytes and acanthocytes. Some fragments were echinocytic and some were acanthocytic. ADAMTS13 was 80%. A diagnosis of transplant-associated thrombotic microangiopathy (TA-TMA) was made, with ciclosporin and bevacizumab as potential precipitants. Following their discontinuation, he was treated with a total of 13 doses of the complement inhibitor eculizumab in addition to defibrotide and dexamethasone. Total hemolytic complement activity (CH50) levels were used to monitor drug activity, and regular blood film and hemolytic marker analysis was used to assess response to therapy and guide treatment duration. He has made a full clinical recovery and at day 230 showed 100% donor chimerism.

TA-TMA is defined by thrombocytopenia and microangiopathic hemolytic anemia with evidence of organ dysfunction – commonly renal failure or neurological impairment. TMA complicates 10–20% of allogeneic stem cell transplants and has a mortality rate of approximately 80%.[1] Endothelial damage appears to be central to the pathophysiology. Etiology of this complex condition in the transplant setting includes intensive conditioning chemotherapy, irradiation, immunosuppressive medication, infection and graft-versus-host disease (GVHD). More recently, a pathophysiological role for terminal complement pathway activation has been recognized.[2]

Hematology: 101 Morphology Updates, First Edition. Barbara J. Bain.
© 2023 John Wiley & Sons Ltd. Published 2023 by John Wiley & Sons Ltd.

Treatment options include replacing calcineurin inhibitors with alternative immunosuppressive agents and the use of rituximab and defibrotide. Plasma exchange is not considered to be effective and is not recommended. Anticomplement therapy with eculizumab has been reported[3] and its use was associated with a good outcome in our patient. This case highlights a number of risk factors for the development of this condition, including GVHD, viral infection and the use of ciclosporin. An association of TMA with bevacizumab has also been recognized.[4] Close monitoring of the peripheral blood film and other blood parameters in patients at risk of TA-TMA is indicated.

Original publication: Renaudon-Smith E, De La Fuente J and Bain BJ (2016) Transplant-associated thrombotic microangiopathy. *Am J Hematol*, **91**, 1160.

References

1 Chapin J, Shore T, Forsberg P, Desman G, Van Besien K and Laurence J (2014) Hematopoietic transplant-associated thrombotic microangiopathy: case report and review of diagnosis and treatments. *Clin Adv Hematol Oncol*, **12**, 565–573.

2 Jodele S, Laskin B, Dandoy CE, Myers KC, El-Bietar J, Davies SM *et al.* (2015) A new paradigm: diagnosis and management of HSCT-associated thrombotic microangiopathy as multi-system endothelial injury. *Blood Rev*, **29**, 191–204.

3 Jodele S, Fukuda T, Vinks A, Mizuno K, Laskin BL, Goebel J *et al.* (2014) Eculizumab therapy in children with severe hematopoietic stem cell transplantation–associated thrombotic microangiopathy. *Biol Blood Marrow Transplant*, **20**, 518–525.

4 Al-Nouri ZL, Reese JA, Terrell DR, Vesely SK and George JN (2015) Drug-induced thrombotic microangiopathy: a systematic review of published reports. *Blood*, **125**, 616–618.

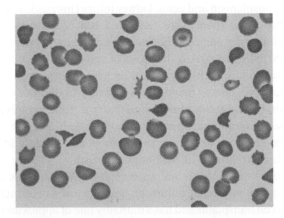

Another image showing schistocytes, including an acanthocytic schistocyte.

50   Neuroblastoma in the bone marrow

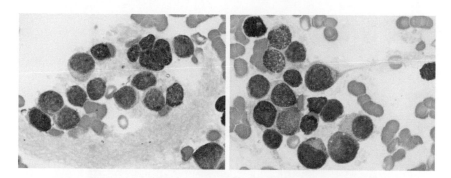

A 4-year-old girl presented to her local hospital with a 3-week history of intermittent fevers, right hip pain and walking with a limp. Blood tests showed raised inflammatory markers (C-reactive protein 203 mg/l, erythrocyte sedimentation rate 118 mm in 1 hour), WBC 12.1 × 10^9/l, Hb 98 g/l, MCV 71 fl, platelets 373 × 10^9/l, neutrophils 6.9 × 10^9/l and lymphocytes 4.1 × 10^9/l. Ultrasound of the right hip did not show any effusion. She was commenced on ceftriaxone and clindamycin for possible osteomyelitis and was transferred to our center for further investigation.

Magnetic resonance imaging of the right hip and pelvis showed a diffuse abnormal bone marrow signal involving all of the bones imaged, suggesting a malignant process involving the bone marrow. The blood count was similar to previously and a blood film showed rouleaux, platelet clumps and 12% large granular lymphocytes. Bone marrow aspiration was difficult and the bone was noted to be abnormally firm. Films of the aspirate showed that normal hemopoietic cells were almost entirely replaced by an infiltrate of neoplastic cells. These were mainly small with a high nucleocytoplasmic ratio and agranular, moderately basophilic cytoplasm. There were some cytological similarities to leukemic blast cells but the correct diagnosis was revealed by the presence of abundant neurofibrillary bundles (left image). Other cytological features indicative of the diagnosis of neuroblastoma were the presence of some much larger cells, the presence of rosettes of neoplastic cells around fibrillar material, the molding of nuclei by the nuclei of adjacent cells (top of the left image) and the presence of occasional cells from which neurofibrils could be seen to emerge (right). It is of interest that nuclear molding is also characteristic of small cell carcinoma of the lung, a tumor that is likewise of neuroendocrine origin. Immunophenotyping confirmed involvement by a non-hematological neoplasm (CD45−, CD56+, CD9+, CD99−). A bone marrow trephine biopsy and immunohistochemistry confirmed the aspirate diagnosis. Computed tomography of the chest and abdomen revealed a calcified mass anterior to the left kidney with enlarged left para-aortic and right axillary lymph nodes, in keeping with the bone marrow diagnosis of neuroblastoma.

A further interesting and potentially significant finding was the fact that the patient was highly sensitive to opiates, with apnea and oxygen desaturation with a relatively low dose of oral morphine (180 µg/kg). Congenital hypoventilation syndrome, the so-called Ondine curse, is associated with a mutation in the *PHOX2B* gene, which is involved in the prenatal development of the autonomic nervous system and is also linked to the development of neuroblastoma. Gene mutational analysis was performed in our patient to investigate this further.

Original publication: Renaudon-Smith E, Arca M, Osei-Yeboah A, Karnik L and Bain BJ (2016) Neuroblastoma in the bone marrow. *Am J Hematol*, **91**, 1272.

Hematology: 101 Morphology Updates, First Edition. Barbara J. Bain.
© 2023 John Wiley & Sons Ltd. Published 2023 by John Wiley & Sons Ltd.

51   Gray platelet syndrome

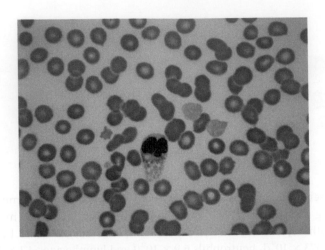

Examination of a blood film is essential when an automated blood count shows thrombocytopenia. Firstly, it is essential to confirm that the platelet count is genuinely low and that the apparent reduction is not attributable to platelet aggregation, satellitism or phagocytosis; the film may also show fibrin strands, indicating partial clotting of the specimen. Secondly, the film may provide a clue to the cause of a congenital or acquired thrombocytopenia when there are small platelets (Wiskott–Aldrich syndrome), large platelets (Bernard–Soulier syndrome, MYH9-related thrombo-cytopenia and others), abnormal platelet granulation (Paris–Trousseau thrombocytopenia) or hypogranular platelets (gray platelet syndrome). Examination of other lineages may also be inform-ative indicating, for example, features of May–Hegglin anomaly, acute promyelocytic leukemia or a thrombotic microangiopathy.

The image shown is from a 20-year-old man who presented with significant hemorrhage into his thigh following a sporting injury. His platelet count was 120×10^9/l. The detection of markedly hypogranular platelets including large and giant platelets led to the diagnosis of gray platelet syndrome, a condition in which α granules, or α and δ granules, are lacking. It is due to biallelic mutation of the *NBEAL2* gene.

Original publication: Bain BJ and Bhavnani M (2011) The gray platelet syndrome. *Am J Hematol*, **86**, 1027.

52   Diagnosis of systemic lupus erythematosus from a bone marrow aspirate

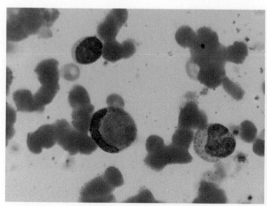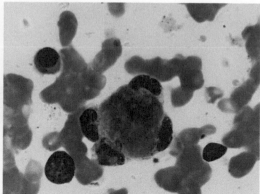

An 18-year-old Iraqi girl presented with fever and a generalized rash. The spleen was palpable 4 cm below the left costal margin. A blood count showed WBC $3 \times 10^9/l$, Hb 97 g/l, platelet count $63 \times 10^9/l$, neutrophil count $2.2 \times 10^9/l$ and lymphocyte count $0.6 \times 10^9/l$. A bone marrow aspirate performed to investigate the thrombocytopenia showed a normocellular marrow with normal numbers of megakaryocytes. In addition, numerous lupus erythematosus (LE) cells were detected (left image). The LE cells were detected in bone marrow films spread immediately without any anticoagulation, incubation or other manipulation. The diagnosis of systemic lupus erythematosus (SLE) was confirmed by the demonstration of antinuclear antibody and anti-double-stranded DNA antibody.

An LE cell is a neutrophil that has ingested the denatured nuclear material of another cell as the result of antinuclear activity in the plasma. This phenomenon was first described in the bone marrow in 1948 with a Feulgen stain being used to demonstrate that the ingested material was DNA.[1] In the following year an *in vitro* test was reported using either buffy coat cells or normal bone marrow components plus the patient's plasma.[2] Hargraves and colleagues[1] also described the phenomenon shown in the right image – "numerous mature polymorphonuclear leukocytes attempting to pick up the same mass of material". With the availability of serological tests, an LE cell preparation is now rarely used for the diagnosis of SLE. However, the observation of LE cells in a bone marrow aspirate or, rarely, in the peripheral blood can still provide confirmation of this diagnosis.

Original publication: Abdulsalam AH, Sabeeh N, Hatim A and Bain BJ (2012) Diagnosis of systemic lupus erythematosus from a bone marrow aspirate. *Am J Hematol*, **87**, 620.

References

1 Hargraves M, Richmond H and Morton R (1948) Presentation of two bone marrow components; the tart cell and the L.E. cell. *Proc Staff Meet Mayo Clin*, **23**, 25–28.
2 Hargraves MM (1949) Production in vitro of the L.E. cell phenomenon; use of normal bone marrow elements and blood plasma from patients with acute disseminated lupus erythematosus. *Proc Staff Meet Mayo Clin*, **24**, 234–237.

53   Diagnosis from a blood film following a dog bite

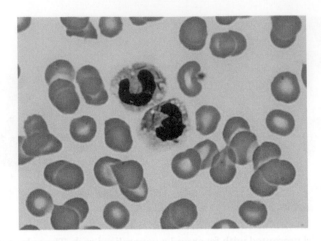

A 60-year-old woman presented in septic shock 2 days after being bitten on the calf by a dog. Twenty-four years previously she had required a splenectomy for autoimmune thrombocytopenic purpura ('ITP') with the platelet count remaining normal thereafter. On clinical examination, the patient had a mottled rash on her left leg which, within 24 hours, had developed into extensive purpura fulminans involving the extremities and the buttocks. She also developed disseminated intravascular coagulation (DIC) and multiorgan failure.

The blood count on admission showed WBC of 11.3×10^9/l, neutrophils 8.3×10^9/l, Hb 113 g/l and platelet count 38×10^9/l. The blood film showed features of hyposplenism and in addition the presence of bacilli within neutrophils (image), supporting a diagnosis of *Capnocytophaga canimorsus* septicemia. This was subsequently confirmed by blood culture. Histology of the excised dog bite site showed necrotic skin with thrombi in small vessels in the underlying subcutaneous tissue. Despite early diagnosis, broad-spectrum antibiotics and intensive support the patient developed gangrene of one hand and the soles of her feet and died 12 days after presentation.

Capnocytophaga canimorsus, formerly known as dysgonic fermenter 2 (DF-2), is a commensal bacterium of dog and cat saliva. It can be transmitted to humans by a bite, a scratch or mere licking. Hyposplenic patients are particularly at risk of septicaemia, DIC and death and should be warned about the risks of close contact with cats and dogs. Urgent examination of a blood film is indicated in such patients presenting with a febrile illness after a dog bite, although even rapid diagnosis and urgent treatment may not prevent death.

Original publication: Tay HS, Mills A and Bain BJ (2012) Diagnosis from a blood film following dog-bite. *Am J Hematol*, **87**, 915.

54 Interpreting a postpartum Kleihauer test

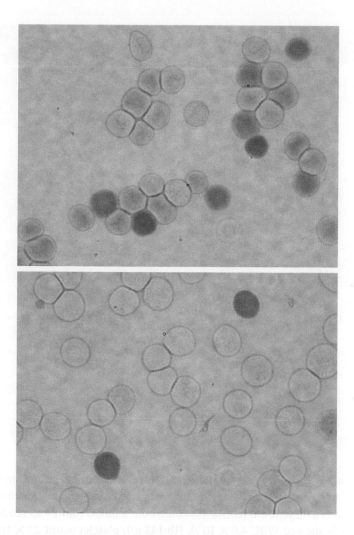

A 23-year-old blood group O RhD-negative woman presented already in labour and gave birth to an O RhD-positive baby. The baby's direct antiglobulin test was negative. Following delivery, the mother received a standard dose of anti-D, and a post-delivery maternal blood sample was sent to quantitate any fetomaternal hemorrhage and thus determine the adequacy of the dose given. The test was positive but it was noted that the distribution of hemoglobin F was heterogeneous (top image) rather than showing the two clear populations of cells expected if a fetomaternal hemorrhage had occurred. The expected pattern is shown in the positive control (bottom). The absence of fetal cells in the maternal circulation was confirmed by flow cytometry, which showed no D-positive cells. High performance liquid chromatography showed the mother to have 4.9% hemoglobin F.

Original publication: Bain BJ and Chapple L (2015) Interpreting a postpartum Kleihauer test. *Am J Hematol*, **90**, 77.

Hematology: 101 Morphology Updates, First Edition. Barbara J. Bain.
© 2023 John Wiley & Sons Ltd. Published 2023 by John Wiley & Sons Ltd.

55   Dengue fever in returning travellers

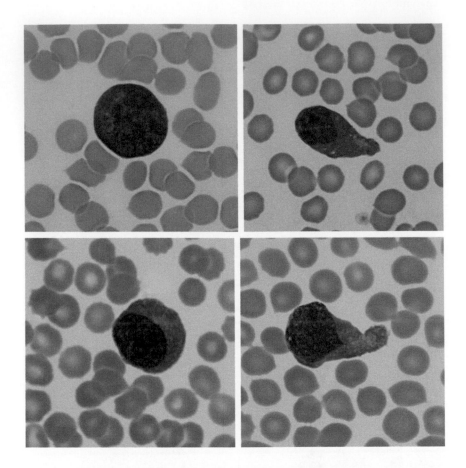

Physicians and hematology laboratories in non-endemic regions need to be alert to the occurrence of tropical diseases in returning travellers. One such imported disease is dengue fever.

The first patient was a 63-year-old febrile man on whom a blood count was requested by a general practitioner. This showed WBC 4.0×10^9/l, Hb 143 g/l, platelet count 22×10^9/l, neutrophil count 2.2×10^9/l and lymphocyte count 1.0×10^9/l. The blood film showed atypical lymphocytes (top images); some of these were plasmacytoid, some resembled immunoblasts and some had a hand-mirror shape. The laboratory staff noticed that the patient's name suggested a Sri Lankan origin and suspected dengue fever. The patient was found to have recently returned from Sri Lanka and the diagnosis was confirmed.

The second patient was a 23-year-old man of Indian ethnic origin who became febrile 2 days after returning from India, where he had been back-packing. His blood count showed WBC 3.2×10^9/l, Hb 164 g/l, platelet count 22×10^9/l, neutrophil count 0.9×10^9/l and lymphocyte count 1.9×10^9/l. His blood film showed moderate numbers of atypical lymphocytes (bottom images); most of these were nucleolated, some were plasmacytoid and hand-mirror forms were frequent. Both laboratory and clinical staff suspected dengue fever and this diagnosis was confirmed.

Hematology: 101 Morphology Updates, First Edition. Barbara J. Bain.
© 2023 John Wiley & Sons Ltd. Published 2023 by John Wiley & Sons Ltd.

In a patient who has visited an endemic area, the combination of severe thrombocytopenia and atypical lymphocytes without lymphocytosis raises the suspicion of dengue fever. The diagnostic value of atypical lymphocytes has been commented on previously.[1] The total white cell count tends to be low normal or reduced and there may also be neutropenia. Patients with more severe disease can have evidence of a consumption coagulopathy. A hemophagocytic syndrome can occur[2] and exceptionally a plasmacytic leukemoid reaction has been reported.[3] Our two cases highlight the importance of assessing both clinical features and blood film morphology.

Original publication: Bain BJ and Stubbs MJ (2015) Dengue fever in returning travellers. *Am J Hematol*, **90**, 263.

References

1 Thisyakorn U, Nimmannitya S, Ningsanond V and Soogarun S (1984) Atypical lymphocyte in dengue hemorrhagic fever: its value in diagnosis. *Southeast Asian J Trop Med Public Health*, **15**, 32–36.
2 Mitra S and Bhattacharyya R (2014) Hemophagocytic syndrome in severe dengue fever: a rare presentation. *Indian J Hematol Blood Transfus*, **30**, Suppl. 1, 97–100.
3 Gawoski JM and Ooi WW (2003) Dengue fever mimicking plasma cell leukemia. *Arch Pathol Lab Med*, **127**, 1026–1027.

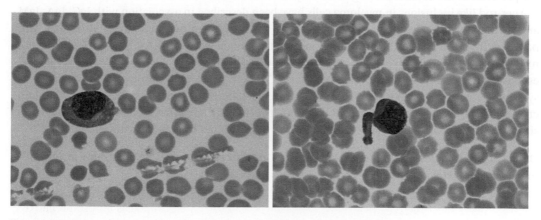

Two further atypical lymphocytes, one plasmacytoid, from the second patient.

56 Auer rod-like inclusions in multiple myeloma

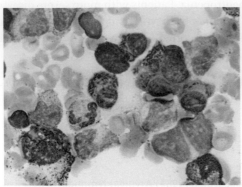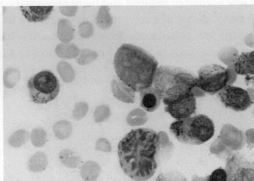

A variety of types of inclusions are seen in the neoplastic cells of multiple myeloma (plasma cell myeloma). Among these, crystalline inclusions resembling Auer rods are uncommon. This 47-year-old Iraqi woman presented with back pain and anemia (Hb 100 g/l). A diagnosis of IgGκ myeloma with Bence Jones proteinuria was made.[1] The bone marrow was heavily infiltrated by cytologically abnormal plasma cells including multinucleated and nucleolated cells (left image). In addition, the majority of neoplastic cells contained azurophilic cytoplasmic crystals. A further curious feature was the presence of mitotic figures within crystal-containing myeloma cells (right), the presence of the crystals serving to identify the cells in mitosis as myeloma cells.

Auer-rod like inclusions in myeloma cells have been described in association with IgG, IgA, IgM and light chain only (kappa) myeloma.[2] They are generally,[2] but not invariably,[3] associated with paraproteins with κ light chains and are more often associated with IgA myeloma. They do not represent immunoglobulin but rather are of lysosomal origin, being positive for α naphthyl acetate esterase, β glucuronidase and acid phosphatase.[4] There is an association with adult Fanconi syndrome due to the deposition of crystals in renal tubules.[2] Myeloma cells can also contain crystalline inclusions that do represent immunoglobulin but these differ from Auer rod-like inclusions in that they stain pale pink on Romanowsky stains, rather than being azurophilic.

Original publication: Abdulsalam AH and Bain BJ (2014) Auer-rod like inclusions in multiple myeloma. *Am J Hematol*, **89**, 338.

References

1 Abdulsalam AH and Al-Yassin FM (2012) Myeloma cells with Auer rod like inclusions. *Turk J Hematol*, **29**, 206.
2 Hütter G, Nowak D, Blau IW and Thiel E (2009) Auer rod-like intracytoplasmic inclusions in multiple myeloma. A case report and review of the literature. *Int J Lab Hematol*, **31**, 236–240.
3 Kulbacki EL and Wang E (2009) IgG-λ plasma cell myeloma with cytoplasmic azurophilic inclusion bodies. *Am J Hematol*, **85**, 516–517.
4 Metzgeroth G, Back W, Maywald O, Schatz M, Willer A, Hehlmann R and Hastka J (2003) Auer rod-like inclusions in multiple myeloma. *Ann Hematol*, **82**, 57–60.

Hematology: 101 Morphology Updates, First Edition. Barbara J. Bain.
© 2023 John Wiley & Sons Ltd. Published 2023 by John Wiley & Sons Ltd.

57   Azurophilic granules in myeloma cells

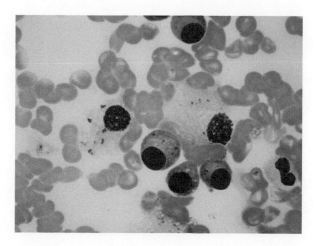

The neoplastic cells of multiple myeloma (plasma cell myeloma) can show a wide range of cytological abnormalities. These include cytoplasmic inclusions composed of immunoglobulin, which may be either crystalline or globular, the latter either clearly in the cytoplasm (Russell bodies) or invaginated into the nucleus (Dutcher bodies). A distinctive type of rod-shaped azurophilic inclusion does not represent immunoglobulin but rather is lysosomal in origin.[1]

In this patient with IgGκ myeloma a large proportion of the myeloma cells contained azurophilic granules, sometimes round and sometimes irregular (image). By analogy with the azurophilic Auer rod-like inclusions with similar staining characteristics that may be observed in myeloma,[1,2] it is likely that these inclusions are of lysosomal origin. In one reported case, the granules were positive for acid phosphatase and weakly positive for α naphthyl butyrate esterase; on ultrastructural examination they were distinct from the endoplasmic reticulum and were considered likely to be vacuoles of the lysosomal system.[3] Azurophilic cytoplasmic granules have been observed in association with both κ[4] and λ[3] light chains.

Original publication: Bain BJ, Siow W, Rahemtulla A and Abdalla S (2014) Azurophilic granules in myeloma cells. *Am J Hematol*, **89**, 437.

References

1 Metzgeroth G, Back W, Maywald O, Schatz M, Willer A, Hehlmann R and Hastka J (2003) Auer rod-like inclusions in multiple myeloma. *Ann Hematol*, **82**, 57–60.
2 Abdulsalam AH and Bain BJ (2014) Auer-rod like inclusions in multiple myeloma. *Am J Hematol*, **89**, 338.
3 Kuyama J, Kosugi S, Take H, Matsuyama T and Kanayama Y (2004) Non-secretory multiple myeloma with azurophilic granules and vacuoles: an immunological and ultrastructural study. *Intern Med*, **43**, 590–594.
4 Vlădăreanu AM, Cîşleanu D, Derveşteanu M, Onisâi M, Bumbea H, Radeşi S *et al.* (2008) Myeloma cells with azurophilic granules – an unusual morphological variant – case presentation. *J Med Life*, **1**, 74–86.

Hematology: 101 Morphology Updates, First Edition. Barbara J. Bain.
© 2023 John Wiley & Sons Ltd. Published 2023 by John Wiley & Sons Ltd.

58   *Plasmodium knowlesi*

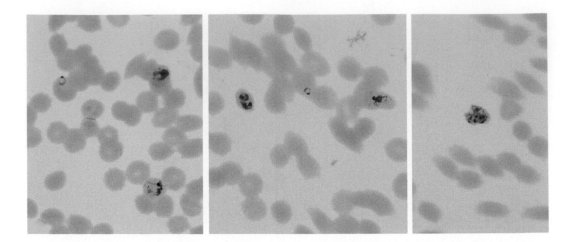

Plasmodium knowlesi has been known as a parasite of South-East Asian monkeys since the first quarter of the last century. It was experimentally transmitted to humans in the 1930s and for a period was used in the treatment of neurosyphilis. Subsequently, naturally occurring human infection was described. In recent decades human cases have been recognized in Indonesia, Borneo, Singapore, Malaysia, Myanmar, Vietnam and the Philippines and occasional cases have occurred elsewhere in returning travellers.

The morphological features share some characteristics with *P. falciparum* in that there are delicate ring forms, sometimes two ring forms per cell and sometimes double chromatin dots and accolé forms (left and center images). The parasitized cells are not enlarged and have fine cytoplasmic dots, which were referred to as Sinton and Mulligan's dots in the older literature. The presence of young rings, accolé forms and high parasitemia may lead to confusion with *P. falciparum*. However, in addition to delicate rings there are also more chunky rings and, later, somewhat ameboid trophozoites (left and center). Schizonts may also be seen with up to 16 merozoites (right), whereas these are quite uncommon in the peripheral blood in *P. falciparum* infection. A common problem is misidentification of *P. knowlesi* as *P. malariae*, especially as both parasites produce band form trophozoites. If a diagnostic laboratory sees what appears to be *P. malariae* in a person who acquired malaria in the Asia-Pacific region, PCR for *P. knowlesi* should be undertaken, especially as *P. knowlesi* produces a much more severe infection and the treatment given should be appropriate.

These images are from *in vitro* cultured parasites kindly provided by Dr Don Van Schalkwyk of the London School of Hygiene and Tropical Medicine.

Original publication: Chiodini P and Bain BJ (2017) *Plasmodium knowlesi. Am J Hematol*, **92**, 716.

59 The cytological features of *NPM1*-mutated acute myeloid leukemia

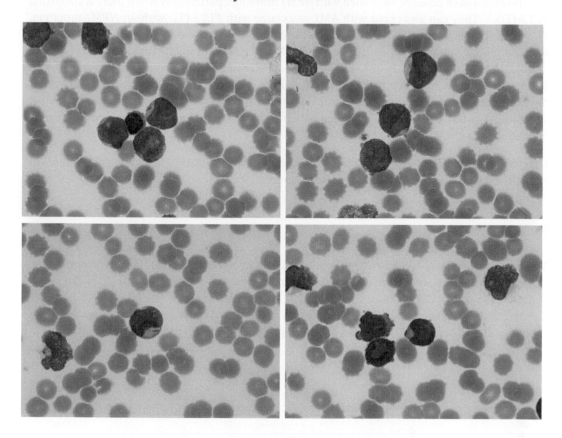

A 69-year-old woman with a previous history of type 2 diabetes mellitus presented with a 2-week history of worsening abdominal pain, reduced oral intake and lethargy. She was pyrexial (39°C) with a sinus tachycardia. Her blood count showed Hb 119 g/l, WBC 151.5 × 10^9/l and platelet count 33 × 10^9/l. The blood film showed many circulating blast cells. These were of medium size with a high nucleocytoplasmic ratio. Many of them had an indentation in the nucleus, producing a cup-like shape when viewed in profile; when viewed from above, the invagination sometimes simulated a giant nuclcolus (images). Nuclear indentations were apparent also in smear cells. Cytoplasmic basophilia varied from moderate to strong. A minority of basophilic blast cells had cytoplasmic blebs. Some blast cells contained Auer rods, usually one but up to three per cell, these often being related to the nuclear indentation. The cytological features were considered strongly suggestive of acute myeloid leukemia (AML) with *NPM1* mutation and further investigations were arranged.

A bone marrow aspirate showed 89% blast cells with similar cytological features to those in the peripheral blood. Flow cytometry gated on the blast population showed expression of CD33, CD117 and CD123 with partial expression of HLA-DR and myeloperoxidase. CD34 was not expressed. Blast cells were also negative for CD13, CD11b, CD14, CD64, CD56 and CD7. Standard cytogenetic analysis yielded no metaphases but molecular analysis showed both a 4 bp insertion in *NPM1* and a *FLT3* internal tandem duplication (ITD) of 24 bp.

Hematology: 101 Morphology Updates, First Edition. Barbara J. Bain.
© 2023 John Wiley & Sons Ltd. Published 2023 by John Wiley & Sons Ltd.

The patient responded to broad-spectrum antibiotics, without a focus of infection being identified, and entered complete remission following the initial course of cytarabine and daunorubicin.

Cup-shaped blast cells are associated with *NPM1* mutation, particularly when there is coexisting *FLT3*-ITD.[1,2] They can also occur with AML associated with *FLT3*-ITD without *NPM1* mutation. The nuclear pockets are shown on ultrastructural examination to be occupied by a collection of mitochondria, lysosomes and endoplasmic reticulum.[1,2] The light microscopy features of *NPM1*-mutated AML are sufficiently distinctive to suggest this diagnosis. It should, however, be noted that not all patients have this distinctive morphology; sometimes the features are those of acute myelomonocytic or monocytic/monoblastic leukemia. The CD34 negativity, often with HLA-DR also being negative, could lead to confusion with the variant form of acute promyelocytic leukemia but the nuclear shape differs.

Original publication: Bain BJ, Heller M, Toma S and Pavlů J (2015) The cytological features of *NPM1*-mutated acute myeloid leukemia. *Am J Hematol*, **90**, 560.

References

1 Chen W, Rassidakis GZ, Li J, Routbort M, Jones D, Kantarjian H *et al.* (2006) High frequency of *NPM1* mutations in acute myeloid leukemia with prominent nuclear invaginations ("cup-like" nuclei). *Blood*, **108**, 1783–1784.
2 Park BG, Chi HS, Jang S, Park CJ, Kim DY, Lee JH *et al.* (2013) Association of cup-like nuclei in blasts with *FLT3* and *NPM1* mutations in acute myeloid leukemia. *Ann Hematol*, **92**, 451–457.

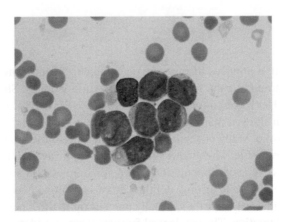

Bone marrow aspirate showing typical cytological features.

60   Irregularly contracted cells in Wilson disease

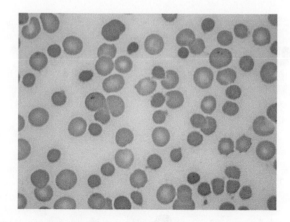

Making a distinction between spherocytes and irregularly contracted cells is important in the diagnostic process. Irregularly contracted cells resemble spherocytes in that they are hyperchromatic and lack central pallor. However, they have an irregular outline and sometimes there are also small protrusions that can be shown to represent Heinz bodies. The above blood film shows polychromatic macrocytes and irregularly contracted cells, some with protrusions. A Heinz body preparation was positive and the cause of the hematological abnormality was found to be liver failure due to Wilson disease. The release of copper from hepatocytes leads to oxidant damage to red cells.

Common causes of irregularly contracted cells are hemolysis in glucose-6-phosphate dehydrogenase deficiency, exposure to oxidant drugs and chemicals and hemoglobinopathies, such as hemoglobin C and unstable hemoglobins.

Original publication: Bain BJ (2008) Irregularly contracted cells. *Am J Hematol*, **83**, 592.

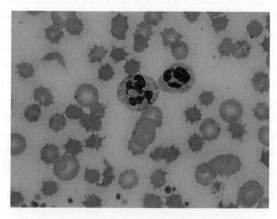

A blood film showing the contrasting features of another type of liver failure, that due to acute on chronic alcohol excess. There are some macrocytes and numerous acanthocytes. This condition is sometimes designated 'spur cell hemolytic anemia'.

Hematology: 101 Morphology Updates, First Edition. Barbara J. Bain.
© 2023 John Wiley & Sons Ltd. Published 2023 by John Wiley & Sons Ltd.

61 Pseudo-Pelger–Huët neutrophil morphology due to sodium valproate toxicity

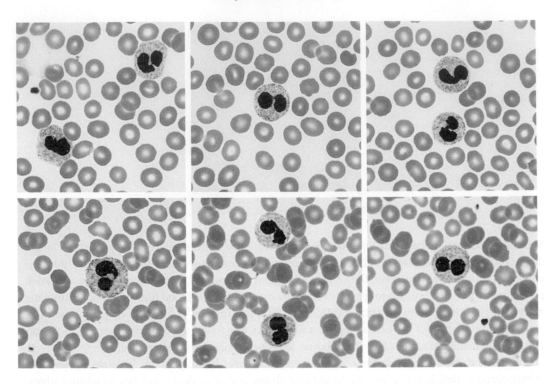

A 57-year-old woman was referred for investigation of chronic thrombocytopenia prior to elective hip replacement. Her blood count showed Hb 129 g/l, MCV 96 fl, WBC 6.5 × 10^9/l, neutrophils 3.9 × 10^9/l, lymphocytes 1.8 × 10^9/l, monocytes 0.7 × 10^9/l and platelets 58 × 10^9/l. She had a history of epilepsy treated with sodium valproate 1 g twice daily. Her blood film confirmed a true thrombocytopenia and was notable for abnormal neutrophil segmentation with frequent bilobed forms and condensed nuclear chromatin, resembling the Pelger–Huët anomaly, with normal cytoplasmic granulation (images). A diagnosis of a myelodysplastic syndrome was considered but it was noted that sodium valproate therapy, particularly in large doses, can cause thrombocytopenia and pseudo-Pelger–Huët neutrophil features. The sodium valproate dose was titrated down and stopped after 4 weeks, being replaced by levetiracetam. The platelet count improved to 156 × 10^9/l and 230 × 10^9/l at 4 and 9 weeks, respectively, and the neutrophil morphology returned to normal.

Pseudo-Pelger–Huët neutrophil morphology can result from exposure to a number of medications including mycophenolate mofetil, tacrolimus, taxols, ganciclovir and valproate. It is important to consider this possibility, particularly in patients with associated cytopenias.

Original publication: Laing A, Harper K, Leach M and Bain BJ (2022) Pseudo-Pelger–Huët neutrophil morphology due to sodium valproate toxicity. *Am J Hematol*, **97**, 1108–1109.

62 The distinctive cytological features of T-cell prolymphocytic leukemia

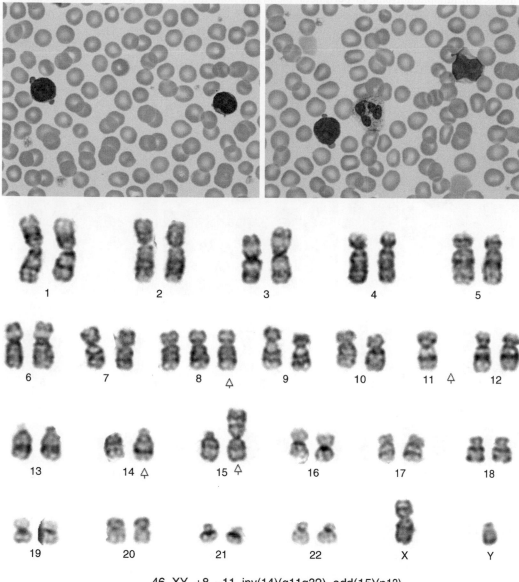

46, XY, +8, −11, inv(14)(q11q32), add(15)(p13)

The first patient was a 64-year-old man, referred for investigation of asymptomatic lymphocytosis. He had no palpable lymphadenopathy or hepatosplenomegaly and no rash. His presenting blood count showed WBC 15.6×10^9/l, Hb 161 g/l, platelet count 165×10^9/l and lymphocytes 10.2×10^9/l. Other than a marginally elevated lactate dehydrogenase of 285 iu/l, biochemical tests were normal. His blood film showed mature lymphocytes with hyperchromatic irregular nuclei, indistinct nucleoli and scanty basophilic cytoplasm forming blebs; in addition there were abnormal lymphoid cells,

Hematology: 101 Morphology Updates, First Edition. Barbara J. Bain.
© 2023 John Wiley & Sons Ltd. Published 2023 by John Wiley & Sons Ltd.

which were larger with a lesser degree of chromatin condensation and with a more evident nucleolus (top images above). A presumptive diagnosis of T-cell prolymphocytic leukemia (T-PLL) was made on the basis of the cytological features. This was confirmed by immunophenotyping and cytogenetic analysis. Immunophenotyping of peripheral blood and bone marrow cells demonstrated an abnormal population of T lymphocytes expressing CD2, CD3, CD4, CD5 and strong CD7 with homogenous coexpression of CD52. They were negative for CD8 and CD25. The karyotype was 46,XY,+8,−11,inv(14)(q11q32),add(15)(p13)[4]/46,idem,?t(2;15)(q21;q2?6)[1]/44,idem,add(2)(q37),−5,−8,−12,add(14)(q32),+19[2]/46,XY[4] (bottom image above).

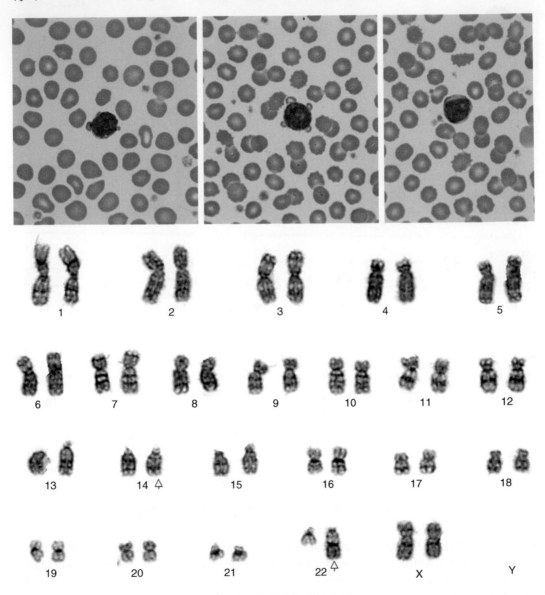

46, XX, inv(14)(q11q32), add(22)(q?11.2)

The second patient, a 72-year-old woman, also had asymptomatic lymphocytosis with no hepatosplenomegaly, lymphadenopathy or rash. Biochemical tests were normal. Her blood count showed WBC $5.9 \times 10^9/l$, Hb 143 g/l, platelet count $333 \times 10^9/l$ and lymphocytes $4.9 \times 10^9/l$. Cytological features (top images above) were similar to those of the first patient but more subtle, again leading to a provisional diagnosis of T-PLL. Peripheral blood immunophenotyping showed that most of the lymphocytes were T cells expressing CD2, CD4 and CD5 with stronger expression of CD3 and CD7 and homogenous expression of CD26, CD28 and CD52. They were negative for CD8, CD11c and CD25. The karyotype was 46,XX,inv(14)(q11q32)[2]/46,idem,add(12)(p1?11.2) [12]/46,idem,t(1;11)(p1;p1)[1]/46,idem,add(22)(q13)[2]/46,XX[3] (bottom image above).

In both patients fluorescence *in situ* hybridization confirmed the inv(14) and showed that the *TRA/TRD* locus was rearranged. Serology for human T-cell lymphotropic virus 1 was negative in both.

The cytological features of T-PLL differ from those of B-PLL. The nucleus is more often hyperchromatic, with the nucleolus being smaller and not so readily detected. The cytoplasm in many of the cells is scanty and basophilic and forms blebs. These features are sufficiently distinctive to strongly suggest this diagnosis. The diagnosis is supported by the detection of strong CD7 expression, this being unusual in other disorders of mature T lymphocytes but typical of T-PLL. Immunophenotyping also serves to demonstrate CD52 expression, which is clinically relevant since alemtuzumab may be used in therapy. In patients with appropriate cytological and immunophenotypic features, the diagnosis is confirmed by the demonstration of a karyotypic abnormality with a 14q32 breakpoint, particularly inv(14)(q11.2q32.1) (as in the current patients) or t(14;14)(q11.2;q32.1); the genes involved are *TCL1A* and *TCL1B* at 14q32.1 and the *TRA/TRD* locus at 14q11.2.

A minority of patients (about 15%) with T-PLL are, like our patients, asymptomatic at diagnosis. This indolent phase can persist for a variable length of time, sometimes several years.[1] However, progression is inevitable and may be very rapid when it occurs.

Original publication: Jayakar V, Cheung K, Yebra-Fernandez E and Bain BJ (2017) The distinctive cytological features of T-cell prolymphocytic leukemia. *Am J Hematol*, **92**, 830–832.

Reference

1 Garand R, Goasguen J, Brizard A, Buisine J, Charpentier A, Claisse JF *et al.* (1998) Indolent course as a relatively frequent presentation in T-prolymphocytic leukaemia: Groupe Francais d'Hematologie Cellulaire. *Br J Haematol*, **103**, 488–494.

63 Eosinophil morphology in the reactive eosinophilia of Hodgkin lymphoma

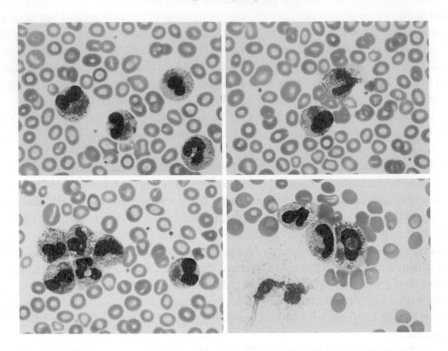

A 35-year-old woman with stage IVB refractory Hodgkin lymphoma, mixed cell type, who had initially presented 18 months earlier, developed marked leukocytosis with striking eosinophilia with disease progression. The eosinophils showed notable morphological abnormalities (images) including degranulation, vacuolation, granules that were smaller than normal and granules in mature eosinophils that had basophilic staining characteristics. There were some hyperlobated nuclei, some non-segmented nuclei (top right) and some ring nuclei (bottom right).

Striking eosinophilia with marked cytological abnormalities often gives rise to the suspicion of eosinophilic leukemia. However, it is important to be aware that marked cytological abnormalities can also occur when eosinophilia is reactive. A systematic study comparing reactive and neoplastic cases reported that, although there was a tendency to more marked morphological abnormalities in those with clonal neoplastic eosinophils, there was considerable overlap.[1] The feature most strongly associated with neoplasia, particularly because of its association with inv(16), was the presence of immature basophilic granules but, as demonstrated in this patient, even this is not specific for neoplasia.

Original publication: Siow W, Matthey F and Bain BJ (2022) Eosinophil morphology in the reactive eosinophilia of Hodgkin lymphoma. *Am J Hematol*, **97**, 373–374.

Reference

1 Goasguen JE, Bennett JM, Bain BJ, Brunning R, Zini G, Vallespi M-T *et al*.; The International Working Group on Morphology of MDS (2020) The role of eosinophil morphology in distinguishing between reactive eosinophilia and eosinophilia as a feature of a myeloid neoplasm. *Br J Haematol*, **191**, 497–504.

64   Malaria pigment

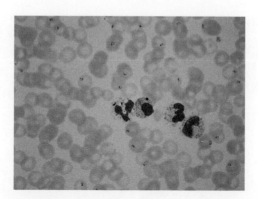

The existence of malaria pigment has been known for well over a century. It was probably first recognized in 1847 by Johann Heinrich Meckel with the connection with malaria being recognized 2 years later by Rudolf Virchow.[1] Dr George Carmichael Low, writing to Dr (later Sir) Patrick Manson, during his expedition to St Lucia in 1901 commented "I went the other day to the dispensary and got five malignant malarias. One child had no symptoms but wasting. Blood contained – Crescents, ring forms, pigmented leucocytes and marked irregularity in size and shape of the corpuscles".[2] Low later suggested that observing the presence of pigmented leucocytes was useful in the differential diagnosis between malignant malaria and yellow fever when malaria parasites were infrequent.[3]

The blood film shown above is from a 54-year-old Afro-Caribbean man who acquired *Plasmodium faliparum* malaria while residing in Uganda. The image demonstrates numerous ring forms of *P. falciparum* plus neutrophils showing toxic granulation and left shift. In addition, two of the neutrophils contain malaria pigment. Malaria pigment is hemozoin or β hematin, an insoluble crystalline derivative of heme. The parasite digests host hemoglobin and converts the heme to hemozoin, thus protecting itself from the toxic effects of heme. A century after Low suggested the diagnostic value of malaria pigment, hemozoin was found to retain diagnostic utility since its ability to depolarize light was found to 'flag' blood samples that were likely to contain malaria parasites.[4]

Original publication: Bain BJ (2011) Malaria pigment. *Am J Hematol*, **86**, 302.

References

1 Sullivan DJ (2004) Hemozoin: a biocrystal synthesized during the degradation of hemoglobin. In: Matsumura S and Steinbüchel A (eds). *Miscellaneous Biopolymers. Biodegradation of Synthetic Polymers*. Wiley-VCH, Weinheim, pp. 130–156.
2 Cook GC (2009) *Caribbean Diseases: Doctor George Low's expedition in 1901–02*. Radcliffe Publishing, Oxford, pp. 58–59.
3 Low GC (1902) The differential diagnosis of yellow fever and malignant malaria. *BMJ*, ii, 860–861.
4 Scott CS, Zyl D, Ho E, Meyersfeld D, Ruivo L, Mendelow BV and Coetzer TL (2003) Automated detection of malaria-associated intraleucocytic haemozoin by Cell-Dyn CD4000 depolarization analysis. *Clin Lab Haematol*, **25**, 77–86.

Hematology: 101 Morphology Updates, First Edition. Barbara J. Bain.
© 2023 John Wiley & Sons Ltd. Published 2023 by John Wiley & Sons Ltd.

65   Salmonella colonies in a bone marrow film

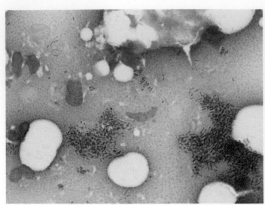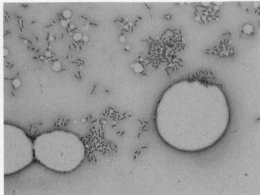

A 27-year-old, previously well, Ghanaian man presented with high fevers, limb pains and a history of diarrhea some days earlier. His spleen was palpable 4 cm below the left costal margin. He was thrombocytopenic (platelet count 26×10^9/l) and mildly anemic (Hb 102 g/l, having been 122 g/l a year previously) with C-reactive protein of 265 mg/l and bilirubin of 83 μmol/l. Cultures were taken and he was started on piperacillin/tazobactam. A computed tomography scan of the chest, abdomen and pelvis showed hepatosplenomegaly (splenic length 17.7 cm) with variegated splenic architecture and bony changes suggesting avascular necrosis in the right humerus and both femoral heads. A day later he was still febrile with progressive pancytopenia (neutrophil count 0.7×10^9/l, having been 3.6×10^9/l on presentation) and was referred to the hematology department. His blood film showed numerous target cells and a few nucleated red blood cells and sickle cells. A bone marrow aspirate showed many necrotic cells and, in addition, numerous bacilli, both single and in clumps (images); contamination was excluded as a possible explanation. Subsequently *Salmonella typhimurium* was demonstrated in the blood culture taken on admission. High performance liquid chromatography showed 47.2% hemoglobin S and 43.1% hemoglobin C.

The patient had a stormy course with cerebral, pulmonary, hepatic and renal impairment. He required prolonged antibiotic therapy, blood transfusion, ventilatory support and hemodialysis, and subsequently draining of a massive splenic abscess, which also grew salmonella; at the time of writing he remained on oral ciprofloxacin as an outpatient, and was well.

Patients with compound heterozygosity for hemoglobins S and C not infrequently present in adult life. As in other forms of sickle cell disease, there is susceptibility to salmonellosis. Infarcted bone marrow provides a rich culture medium for micro-organisms but it is extraordinary for such a large number of organisms to be visualized in the bone marrow, and for the patient to recover from such a severe infection.

Original publication: Ghara N, Portsmore S, Yardumian A and Bain BJ (2014) Salmonella colonies in a bone marrow film. *Am J Hematol*, **89**, 780.

66   Severe babesiosis due to *Babesia divergens* acquired in the UK

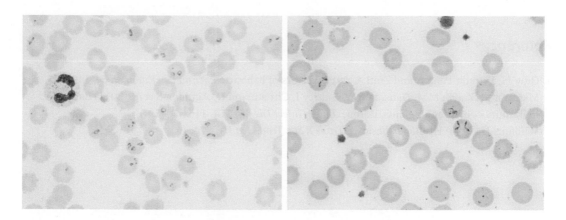

A 72-year-old woman, resident in the UK, presented to her local emergency department with a 3-day history of feeling non-specifically unwell and febrile, with nausea, abdominal pain and dark urine. She had an intact spleen and an unremarkable medical history. Her initial blood count showed Hb 75 g/l with a normal MCV, platelet count 69×10^9/l and normal white cell count and differential. The thrombocytopenia prompted examination of a blood film which showed echinocytosis, true thrombocytopenia and platelet anisocytosis. Neutrophils appeared mildly left-shifted with heavy granulation. More than 20% of red cells contained small ring trophozoites (left image, Field stain). Pyriform pairs were seen (right image, Giemsa stain) and some red cells contained up to six parasites. Pathognomonic tetrads (Maltese cross form), formed of four pyriform parasites joined at their pointed ends, were seen and recognized as morphologically diagnostic of babesiosis. Very infrequent extracellular parasites were seen. PCR confirmed the species as *Babesia divergens*. This species resides in cattle in which it is prevalent in Western and Central Europe, unlike *B. microti*, which is associated with the northeast and upper midwest USA and is well documented as a cause of human babesiosis.

This report highlights the morphology of *B. divergens* infection in humans, a very rare illness, with approximately 50 reported cases, including a previous report of infection in a person with an intact spleen.[1,2]

Babesia divergens parasites are 0.4 × 1.5 μm in size. Whilst they are seen classically as paired forms, diverging up to 180°, on the periphery of the red cell, they are pleomorphic and their size can vary depending upon the host they infect. In fulminant human cases, *B. divergens* takes the form of rings, loops, clubs, rods and pyriform and amoeboid shapes. There can be one to eight parasites per red cell. Parasitemia as high as 70% has been reported from a fatal case.[3] The Maltese cross form is unique to *Babesia* among members of the Apicomplexa, but if not present it may be very difficult to distinguish *Babesia* from young ring forms of *Plasmodium* spp., especially *P. falciparum*. The absence of pigment cannot be relied upon, as young rings of *Plasmodium* spp. do not exhibit pigment. *Babesia* is smaller than malaria parasites, and in some of the larger rings there is a white vacuole, instead of the pink vacuole containing erythrocyte stroma seen in malaria. *Babesia* does not form schizonts and malaria parasites do not form pyriform pairs or tetrads.

Hematology: 101 Morphology Updates, First Edition. Barbara J. Bain.
© 2023 John Wiley & Sons Ltd. Published 2023 by John Wiley & Sons Ltd.

Original publication: Chan WY, MacDonald C, Keenan A, Xu K, Bain BJ and Chiodini PL (2021) Severe babesiosis due to *Babesia divergens* acquired in the United Kingdom. *Am J Hematol*, **96**, 889–890.

References

1 Gray JS, Estrada-Peña A and Zintl A (2019) Vectors of babesiosis. *Annu Rev Entomol*, **64**, 149–165.
2 Martinot M, Zadeh MM, Hansmann Y, Grawey I, Christmann D, Aguillon S *et al.* (2011) Babesiosis in immunocompetent patients, Europe. *Emerg Infect Dis*, **17**, 114–116.
3 Williams H (1980) Human babesiosis. *Trans R Soc Trop Med Hyg*, **74**, 157.

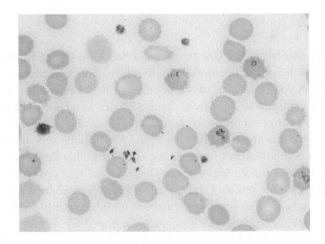

A further image showing intracellular parasites and a cluster of the uncommon extracellular parasites.

67 Congenital acute megakaryoblastic leukemia

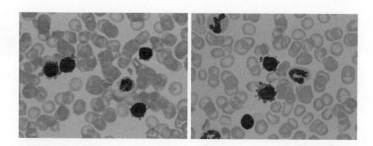

A newborn baby girl, born to non-consanguineous Northern European parents, presented with an epistaxis at 10 days of age. She was found to have hepatosplenomegaly. There were no clinical features of Down syndrome. Her blood count showed WBC 36×10^9/l, Hb 133 g/l and platelet count 74×10^9/l. Her peripheral blood film (images) showed numerous blast cells with distinctive features. They had a high nucleocytoplasmic ratio and small indistinct nucleoli. The scanty cytoplasm varied from weakly to moderately basophilic and extended into irregular basophilic blebs and fronds. There were also some giant platelets. The cytological features were those of acute megakaryoblastic leukemia.

Bone marrow aspiration was difficult. The hemodilute, aparticulate aspirate showed 18% blast cells. Cytogenetic analysis of aspirated bone marrow showed t(1;22)(p13;q13). This subtype of acute myeloid leukemia is associated with an *RBM15::MKL1* fusion gene. It usually presents within the first 6 months of life and at least four congenital cases have been reported previously, in addition to a previous brief report of this case.[1–3] Its occurrence in identical twins at the age of 2 months provides further evidence of an intrauterine origin.[4] The bone marrow blast cell percentage may be below 20%.

Following several courses of combination chemotherapy, remission was achieved but the bone marrow remained hypocellular with low peripheral blood counts. Following sibling hematopoietic stem cell transplantation, she achieved long-term relapse-free survival.

Original publication: Bain BJ, Chakravorty S and Ancliff P (2015) Congenital acute megakaryoblastic leukemia. *Am J Hematol*, **90**, 963.

References

1 Sait SN, Brecher ML, Green SN and Sandberg AA (1988) Translocation t(1;22) in congenital acute megakaryocytic leukemia. *Cancer Genet Cytogenet*, **34**, 277–280.

2 Duchayne E, Fenneteau O, Pages MP, Sainty D, Arnoulet C, Dastugue N *et al.*; Groupe Français d'Hématologie Cellulaire; Groupe Français de Cytogénétique Hématologique (2003) Acute megakaryoblastic leukaemia: a national clinical and biological study of 53 adult and childhood cases by the Groupe Français d'Hématologie Cellulaire (GFHC). *Leuk Lymphoma*, **44**, 49–58.

3 Bain BJ, Murray JA, Patterson KG, Chakravorty S, Ancliff P, Wong CC *et al.* (2005) Slide session, British Society for Haematology, 45th Annual Scientific Meeting, Manchester, 2005. *Clin Lab Haematol*, **27**, 363–369.

4 Ng KC, Tan AM, Chong YY, Lau LC and Lu J (2002) Congenital acute megakaryoblastic leukaemia (M7) with chromosomal t(1;22)(p13;q13) translocation in a set of identical twins. *J Pediatric Hematol Oncol*, **21**, 428–430.

Hematology: 101 Morphology Updates, First Edition. Barbara J. Bain.
© 2023 John Wiley & Sons Ltd. Published 2023 by John Wiley & Sons Ltd.

68 Basophilic differentiation in transient abnormal myelopoiesis

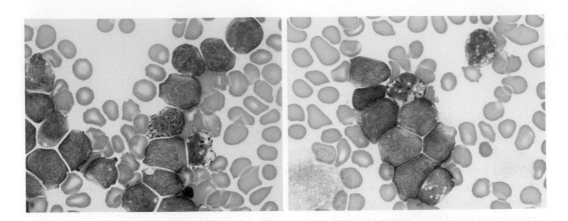

A term baby was born with hydrops fetalis attributable to severe anemia. The blood count showed WBC 285×10^9/l, Hb 70 g/l, MCV 103 fl and platelet count 87×10^9/l. The blood film showed 93% blast cells. There was an increase in the absolute neutrophil count with the neutrophils being cytologically normal. Notable was an increase in the basophil count with mature but cytologically abnormal basophils constituting 2% of cells (images); many of the basophils were hypogranular and vacuolated. The blast cells were medium sized with a high nucleocytoplasmic ratio and moderate cytoplasmic basophilia; some were vacuolated and some contained granules with the same staining characteristics as those in the mature basophils. There were also some nucleated red blood cells and occasional circulating micromegakaryocytes. Down syndrome was suspected and was confirmed by fluorescence *in situ* hybridization which demonstrated trisomy 21. Immunophenotyping showed that the blast cells expressed CD13, CD33, CD34, CD38, CD105 and CD117. There was aberrant expression of CD7.

Differentiation in transient abnormal myelopoiesis of Down syndrome is variable from case to case. Often multiple lineages are involved. Megakaryocytic differentiation is often prominent and sometimes also erythroid. This baby was unusual in having prominent basophilic as well as neutrophilic differentiation.

Original publication: Bain BJ (2016) Basophilic differentiation in transient abnormal myelopoiesis. *Am J Hematol*, **91**, 847.

69 Methylene blue-induced Heinz body hemolytic anemia in a premature neonate

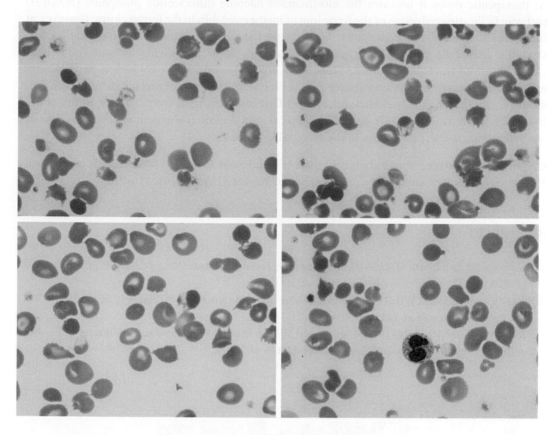

A baby of Northern European ancestry, was born by uncomplicated vaginal delivery at 33 weeks' gestation following induction because of abnormal umbilical artery Dopplers. He was known to have trisomy 21, complicated by duodenal atresia, but was otherwise healthy at birth. At day 2 of life he underwent a duodeno-duodenostomy utilizing methylene blue for intraoperative confirmation of anastomotic integrity. A higher than recommended dose was used leading to classic green/ blue urine and an inability to detect oxygen saturation on pulse oximetry as a result of skin discoloration. He became significantly jaundiced over the next few days (bilirubin 404 μmol/l) with a mixed hyperbilirubinemia (unconjugated 197 μmol/l). His blood film showed marked oxidative hemolytic changes including irregularly contracted cells, blister cells and ghost cells (images). Heinz bodies were visible within ghost cells and blister cells (top left and right), this being confirmed on supravital staining. His Hb dropped from 192 to 67 g/l necessitating top-up blood transfusion. Phototherapy was initially commenced but stopped due to the risk of bronzing in the presence of conjugated hyperbilirubinaemia. The baby recovered fully from the hemolytic crisis without skin blistering or desquamation. Glucose-6-phosphate dehydrogenase (G6PD), checked on a pretransfusion sample, was normal.

Hematology: 101 Morphology Updates, First Edition. Barbara J. Bain.
© 2023 John Wiley & Sons Ltd. Published 2023 by John Wiley & Sons Ltd.

Methylene blue has many uses including checking anastomotic integrity, determining tissue dysplasia (chromoendoscopy) and treatment of ifosfamide neurotoxicity and methemoglobinemia.[1] At therapeutic doses it activates the nicotinamide adenine dinucleotide phosphate (NADPH) pathway facilitating reduction of the ferric iron of methemoglobin to the ferrous form. However, at high doses it behaves as an oxidant and is therefore contraindicated in patients with G6PD deficiency. Oxidative hemolytic effects of methylene blue at high doses in patients without G6PD deficiency have been previously reported.[2,3] Toxicity includes skin and urine discoloration, phototoxicity and respiratory distress. Additionally, methylene blue at toxic doses can precipitate serotonin syndrome due to it monoamine oxidase inhibition.[1] This case highlights the diagnostic utility of the blood film when combined with appropriate clinical information. It also demonstrates the risks of methylene blue usage, particularly in the susceptible neonate.

Original publication: Vanhinsbergh L, Uthaya S and Bain BJ (2018) Methylene blue-induced Heinz body hemolytic anemia in a premature neonate. *Am J Hematol*, **93**, 716–717.

References

1 Ginimuge PR and Jyothi SD (2010) Methylene blue: revisited. *J Anaesthesiol Clin Pharm*, **26**, 517–520.
2 Sills MR and Zinkham WH (1994) Methylene blue induced Heinz body haemolytic anaemia. *Arch Pediatr Adolesc Med*, **148**, 306–310.
3 Vincer MJ, Allen AC, Evans JR, Nwaesei C and Stinson DA (1987) Methylene-blue-induced hemolytic anemia in a neonate. *Can Med Assoc J*, **136**, 503–504.

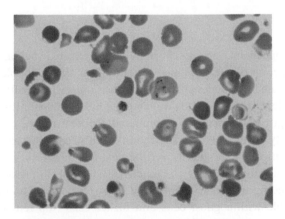

A further image of the baby's blood film showing schistocytes, irregularly contracted cells, a hemighost and a cell containing Pappenheimer bodies.

70   Neutrophil vacuolation in acetominophen-induced acute liver failure

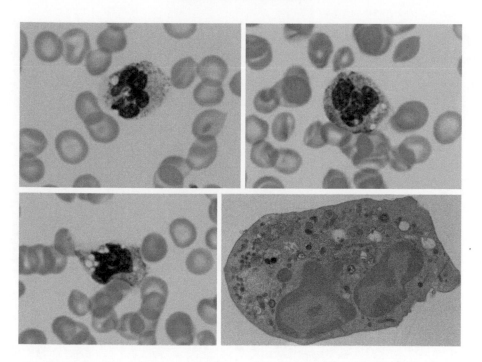

An 18-year-old female was admitted to the liver intensive care unit with hyperacute liver failure in February 2012, 4 days after taking a staggered overdose of 15 g of acetaminophen. The patient had initially presented to her local hospital 24 hours earlier with abdominal pain and blood tests had then revealed an acetaminophen level of 65 mg/l, alanine aminotransferase 6228 iu/l, aspartate aminotransferase 5600 iu/l, bilirubin 51 μmol/l, international normalized ratio 3.3, lactate 4.4 mmol/l and creatinine 48 μmol/l. She had no past medical history of note, took no regular medications and did not drink alcohol. She developed grade 3 hepatic encephalopathy within the first hour of admission with cardiovascular collapse, metabolic acidosis and renal failure. Her blood count showed WBC 15.6×10^9/l, neutrophils 13.9×10^9/l, Hb 137 g/l and platelets 209×10^9/l. Clinically, there was no evidence of infection, C-reactive protein was <2 mg/l and blood and urine cultures were negative. Plasma interleukin (IL)-6, IL-8 and granulocyte colony-stimulating factor (G-CSF) were markedly increased. A blood film revealed neutrophils with marked toxic granulation and prominent vacuolation (images); vacuolation was also apparent on electron microscopy (bottom right). The patient was commenced on an acute liver failure management pathway and was listed for emergency liver transplantation. She underwent a right lobe auxillary liver transplant 5 days after her initial presentation.

Neutrophil vacuolation is usually indicative of bacterial infection and is otherwise uncommon. We have observed it in patients with acute liver failure who are free of infection. We postulate that it is the result of increased levels of G-CSF.

Original publication: Vijay GKM, Kronsten VT, Bain BJ and Shawcross DL (2015) Neutrophil vacuolation in acetaminophen-induced acute liver failure. *Am J Hematol*, **90**, 461.

Hematology: 101 Morphology Updates, First Edition. Barbara J. Bain.
© 2023 John Wiley & Sons Ltd. Published 2023 by John Wiley & Sons Ltd.

71 Howell–Jolly bodies in acute hemolytic anemia

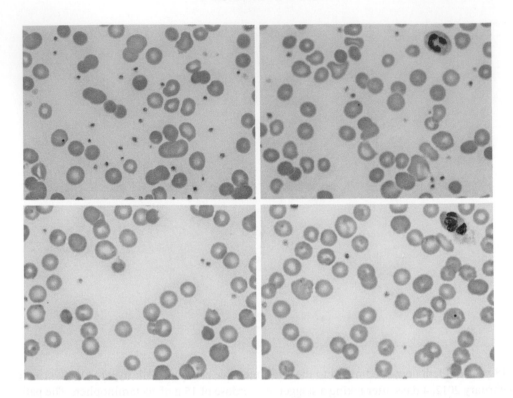

Howell–Jolly bodies are not infrequently observed is acute hemolytic anemia when the hemolysis is largely extravascular.

The upper images show the blood film of a 62-year-old woman with acute warm autoimmune hemolytic anemia. Her Hb was 68 g/l, MCV 110 fl and MCHC 306 g/l. Her blood film showed spherocytes, polychromatic macrocytes and quite frequent Howell–Jolly bodies. A direct antiglobulin test showed immunoglobulin +++++ and C3d +++. The autoantibody showed no specificity. The raised MCV and reduced MCHC were the result of marked reticulocytosis; reticulocytes are not only much larger than mature red cells but also have a considerably lower hemoglobin concentration.

The lower images show the blood film of a 7-year-old boy with acute hemolysis in glucose-6-phosphate dehydrogenase deficiency with an Hb of 58 g/l. His film showed irregularly contracted cells, keratocytes, polychromatic macrocytes, nucleated red blood cells, Pappenheimer bodies and quite frequent Howell–Jolly bodies.

It is postulated that the presence of significant numbers of Howell–Jolly bodies in patients with acute hemolytic anemia is indicative of splenic macrophages that are overloaded with phagocytosed red cells and cellular debris and are thus unable to carry out their normal 'pitting' function. The Pappenheimer bodies are likely to result from the same mechanism.

Original publication: Bain BJ (2017) Howell–Jolly bodies in acute hemolytic anemia. *Am J Hematol*, **92**, 473.

Hematology: 101 Morphology Updates, First Edition. Barbara J. Bain.
© 2023 John Wiley & Sons Ltd. Published 2023 by John Wiley & Sons Ltd.

72 The distinctive micromegakaryocytes of transformed chronic myeloid leukemia

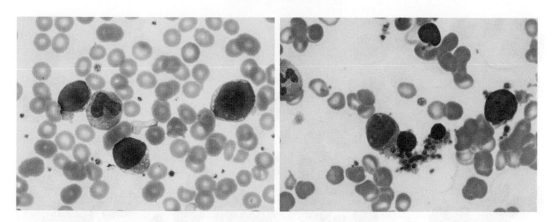

A 41-year-old Afro-Caribbean woman presented with a month's history of worsening abdominal pain. She was pale and had massive splenomegaly. Her blood count showed WBC 66.6×10^9/l, Hb 80 g/l and platelet count of 354×10^9/l. The differential count showed neutrophils 16.8×10^9/l, myelocytes 4×10^9/l, lymphocytes 3.8×10^9/l, eosinophils 9.4×10^9/l, basophils 8.1×10^9/l and blast cells 24.5×10^9/l. The circulating blast cells had weakly basophilic cytoplasm, with some cytoplasmic vacuolation and budding; there were occasional circulating micromegakaryocytes with more condensed chromatin and delicately granular cytoplasm resembling that of a platelet (left image). Because of the characteristic differential count, transformed chronic myeloid leukemia was suspected.

A bone marrow aspirate showed active granulopoiesis. There were 34% blast cells, which were cytologically similar to those in the peripheral blood. Micromegakaryocytes constituted 24% of nucleated cells. Many of these were very distinctive, being lymphocyte-sized with condensed chromatin and with cytoplasm fragmenting into platelets (right image). Cytogenetic analysis showed 46,XX,t(9;22)(q34;q11.2)[3]/46,XX,add(7)(p15),t(9;22)(q34;q11.2)[18]/46,XX[1], confirming the diagnosis of transformed chronic myeloid leukemia with clonal evolution. Fluorescence *in situ* hybridization (FISH) using a Vysis (Abbott) dual fusion probe found 100% of the cells scored to have *BCR::ABL1* and *ABL1::BCR* fusions, with 20% having duplicated fusion signals. Immunophenotyping showed the blast cells to be negative for myeloperoxidase.

In the chronic phase of chronic myeloid leukemia, megakaryocytes are smaller than normal but micromegakaryocytes are not a feature. However, when megakaryoblastic/megakaryocytic transformation occurs, micromegakaryocytes are often numerous and lymphocyte-sized micromegakaryocytes with active platelet production are particularly characteristic.

Original publication: Fordham NJ, Stern S, Swansbury J and Bain BJ (2016) The distinctive micromegakaryocytes of transformed chronic myeloid leukemia. *Am J Hematol*, **91**, 350.

73 Copper deficiency

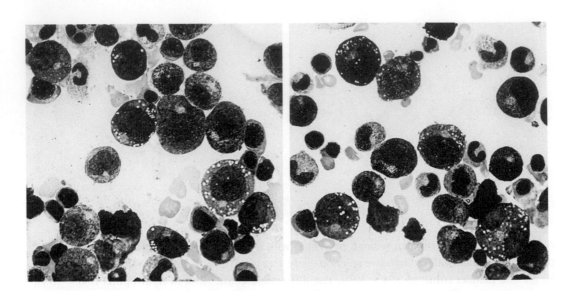

A 44-year-old man with cerebral palsy and epilepsy required a jejunal feeding tube due to recurrent episodes of aspiration pneumonia. He developed progressive anemia and neutropenia – Hb 60 g/l, MCV 85 fl, WBC 1.6×10^9/l, neutrophils 0.5×10^9/l and platelets 165×10^9/l. The blood film showed mild hypochromia and occasional target cells.

The bone marrow aspirate showed a partial myeloid maturation arrest. In addition, there was marked vacuolation of myeloid precursors and proerythroblasts (images) typical of copper deficiency. His serum copper level was 0.2 µmol/l (normal range 10–20). Iron stores were absent and no ring sideroblasts or hemosiderin inclusions in plasma cells were detected. He was commenced on copper supplements and the blood count normalized within 4 weeks – Hb 151 g/l, WBC 8.3×10^9/l and platelets 260×10^9/l.

Copper is required for the function of a number of metabolic co-enzymes, including those involved in iron utilization for hemoglobin synthesis. Copper deficiency can occur in patients on enteral feeding, following bariatric surgery, in malabsorption syndromes, as a result of renal loss of caeruloplasmin in severe nephrotic syndrome, due to excess chelation in Wilson disease and also due to zinc or silver toxicity, the latter following ill-advised intravenous colloidal silver infusion. Copper deficiency typically causes anemia (microcytic, normocytic or macrocytic) and neutropenia with preservation of the platelet count but both thrombocytopenia and pancytopenia can also occur. Some patients show a coexistent myelopathy or peripheral neuropathy, which may not fully recover following supplementation. It is an important condition to recognize and can be mistaken for a myelodysplastic syndrome as the block in iron utilization generates ring sideroblasts in some patients.

Original publication: Al-Bubseree B, Leach M, Jones R and Bain BJ (2020) The hematological effects of copper deficiency. *Am J Hematol*, **95**, 446.

Hematology: 101 Morphology Updates, First Edition. Barbara J. Bain.
© 2023 John Wiley & Sons Ltd. Published 2023 by John Wiley & Sons Ltd.

74 Chronic neutrophilic leukemia

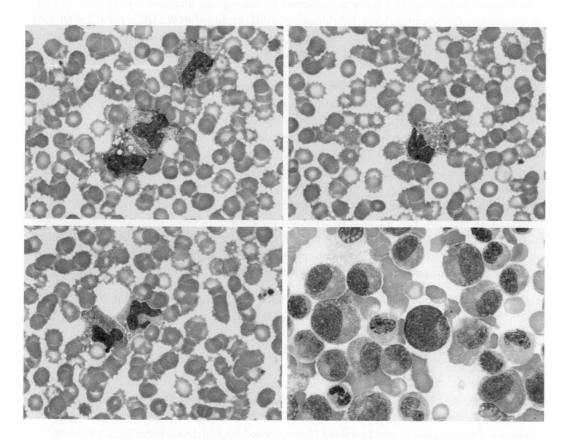

An 83-year-old man with a past medical history of chronic kidney disease, hypertension, ischemic heart disease, atrial fibrillation, previous cerebrovascular accident, hypercholesterolemia and gout was undergoing surveillance after an excised colonic polyp showed adenocarcinoma. He was hospitalized with anemia, sudden worsening of chronic leukocytosis and deteriorating renal function. At that time he had hematuria, hemoptysis and epistaxis. There was no splenomegaly on physical examination. Initial investigations showed normocytic anemia (Hb 70 g/l and MCV 86 fl), WBC 50×10^9/l, platelet count 139×10^9/l and neutrophil count 45×10^9/l. The neutrophils showed toxic granulation and occasional Döhle bodies (top and bottom left images). Granulocyte precursors were less than 10% and monocytes were virtually absent. Serum vitamin B_{12} was greatly elevated (>2000 ng/l) and lactate dehydrogenase was also raised. The patient had no signs of active infection or inflammation and the differential diagnosis was therefore considered to be chronic neutrophilic leukemia (CNL) or a leukemoid reaction, such as that seen with plasma cell neoplasms (multiple myeloma or monoclonal gammopathy of undetermined significance). A bone marrow aspirate (bottom right image) was markedly hypercellular with left-shifted granulopoiesis but no increase of blast cells. Erythropoiesis was greatly reduced. Plasma cells were cytologically normal and not increased in number. Stainable iron was increased but no ring sideroblasts were seen. There was a mild balanced increase in serum free light chains and no paraprotein was

Hematology: 101 Morphology Updates, First Edition. Barbara J. Bain.
© 2023 John Wiley & Sons Ltd. Published 2023 by John Wiley & Sons Ltd.

detected. Computed tomography scanning of the chest and abdomen disclosed no splenomegaly, lymphadenopathy, bone lesions or signs of a neoplastic condition.

The suspected diagnosis of CNL was confirmed by subsequent analyses. Cytogenetic analysis was normal and fluorescence *in situ* hybridization (FISH) excluded *BCR::ABL1*, *FIP1L1::PDGFRA* and rearrangement of *PDGFRB* and *FGFR1*. Sequencing of a panel of 33 genes, known to be pathogenic in myeloid neoplasms, detected a T618I mutation in *CSF3R* exon 14.[1] No mutations in *SETBP1* or *ASXL1* were detected.

The patient was keen to have a trial of treatment but was not responsive to a renally-adjusted dose of hydroxycarbamide or subsequent cyclophosphamide or idarubicin plus etoposide. The WBC peaked at 198×10^9/l with death occurring within 2 months from presentation.

The molecular abnormality most strongly associated with CNL is a *CSF3R* mutation with a minority of patients having a *JAK2* V617F mutation. Interestingly, the blood film is not particularly useful in distinguishing *CSF3R*-mutated CNL from a leukemoid reaction since toxic granulation and Döhle bodies are common in both,[2] in one instance resulting from the physiological action of granulocyte colony-stimulating factor (G-CSF) and in the other from a mutation in the gene encoding the G-CSF receptor. Investigation to exclude a plasma cell or other neoplasm and molecular analysis to confirm CNL are therefore essential. The minority of cases with a *JAK2* V617F mutation less often have 'toxic' changes in neutrophils, this being reported in only two of eight published cases, when the publication permitted this to be assessed.[1]

Original publication: Khoder A, Al Obaidi M, Babb A, Liu C, Cross NCP and Bain BJ (2018) Chronic neutrophilic leukemia. *Am J Hematol*, **93**, 841–842.

References

1 Bain BJ, Brunning RD, Orazi A and Thiele J (2017) Chronic neutrophilic leukaemia. In: Swerdlow SH, Campo E, Harris NL, Jaffe ES, Pileri S, Stein H and Thiele J (eds). *WHO Classification of Tumours of Haematopoietic and Lymphoid Tissues*, revised 4th edn. IARC Press, Lyon, pp. 37–38.

2 Bain BJ and Ahmad S (2015) Chronic neutrophilic leukaemia and plasma cell-related neutrophilic leukaemoid reactions. *Br J Haematol*, **171**, 400–410.

75 Neutrophilic leukemoid reaction in multiple myeloma

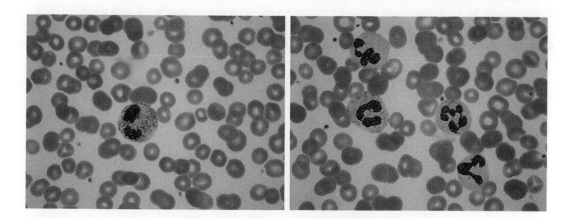

A 52-year-old man with asymptomatic multiple myeloma was referred for assessment of a high white cell count and neutrophil count. He had already been commenced on hydroxycarbamide and while on this drug in a dose of 2.5 g daily, his WBC was 22.5×10^9/l and neutrophil count 18.9×10^9/l. His blood film showed toxic granulation and Döhle bodies (images), although not all neutrophils were abnormal. He was found to have an IgGλ paraprotein in a concentration of 80 g/l. A bone marrow aspirate showed 10–15% plasma cells and granulocytic hyperplasia. Molecular analysis for *JAK2* V617F and *CSF3R* exon 14 and 17 mutations was negative. A diagnosis of neutrophilic leukemoid reaction to multiple myeloma was made.

Isolated neutrophilia without dysplasia is characteristic of chronic neutrophilic leukemia (CNL). However, before making this diagnosis, it is essential to exclude an underlying plasma cell neoplasm. About a quarter of the reported patients with either CNL or a condition resembling it, have had coexisting monoclonal gammopathy of undetermined significance or multiple myeloma.[1] A diagnosis of CNL is not appropriate in patients with a coexisting plasma cell neoplasm unless there is demonstrated clonality of myeloid cells or other clear evidence that the myeloid condition is neoplastic. Rarely, by chance, CNL and a plasma cell neoplasm have coexisted.[2,3] Patients with a plasma cell neoplasm and neutrophilia often have toxic granulation and Döhle bodies, as shown in the current patient. Synthesis of granulocyte colony-stimulating factor (G-CSF) by neoplastic plasma cells and high levels of serum G-CSF have been demonstrated, indicating that this is a reactive, cytokine-driven process. Although similar changes are also seen in CNL, in these patients the cytological changes are likely to result from a mutation that directly affects neutrophil biology, specifically a mutation of *CSFR3*, encoding the G-CSF receptor.[4]

Original publication: Milojkovic D, Hunter A, Barton L, Cross NCP and Bain BJ (2015) Neutrophilic leukemoid reaction in multiple myeloma. *Am J Hematol*, **90**, 1090.

References

1 Bain BJ and Ahmad S (2015) Chronic neutrophilic leukaemia and plasma cell-related neutrophilic leukaemoid reactions. *Br J Haematol*, **171**, 400–410.

2 Blombery P, Kothari J, Yong K, Allen C, Gale RE and Khwaja A (2014) Plasma cell neoplasm associated chronic neutrophilic leukemia with membrane proximal and truncating *CSF3R* mutations. *Leuk Lymphoma*, **55**, 1661–1662.

3 Nedeljkovic M, He S, Szer J and Juneja S (2014) Chronic neutrophilia associated with myeloma: is it clonal? *Leuk Lymphoma*, **55**, 439–440.

4 Maxson JE, Gotlib J, Pollyea DA, Fleischman AG, Agarwal A, Eide CA *et al.* (2013) Oncogenic CSF3R mutations in chronic neutrophilic leukemia and atypical CML. *New Engl J Med*, **368**, 1781–1790.

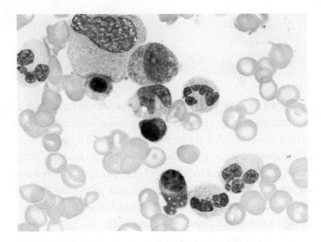

The patient's bone marrow showing a vacuolated neutrophil and a plasma cell with round cytoplasmic inclusions.

76 Persistent polyclonal B lymphocytosis

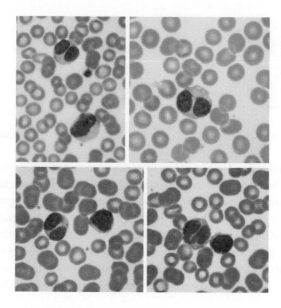

A 37-year-old man presented with abdominal pain. On examination, his spleen was enlarged 2 cm below the left costal margin but there was no superficial lymphadenopathy. He denied any weight loss or B symptoms. His blood count showed WBC 18.6×10^9/l, Hb 154 g/l, platelet count 181×10^9/l and lymphocyte count 13.1×10^9/l. Examination of his blood film permitted a diagnosis to be made, this being subsequently confirmed by flow cytometric immunophenotyping. His blood film showed binucleated lymphocytes (top left and bottom left images). In addition, there were cells that appeared to be tetraploid but in which two nuclear lobes, each the same size as a normal lymphocyte nucleus, were joined by a very narrow isthmus (top right). Other cells had apparently tetraploid, deeply cleft nuclei (bottom right) and others were increased in size with features of immaturity, such as reduced chromatin clumping or the presence of a nucleolus (lower cell, top left).

Immunophenotyping showed that B cells accounted for more than half of the total nucleated cells. They expressed CD19, CD20, CD22, CD79b and CD200 but not CD5, CD10 or CD11c; FMC7 and CD123 were expressed on only a minority of cells. Lymphocytes were polytypic (kappa : lambda 1.5 : 1). Around 6% had a higher forward scatter (indicating larger cells) and these were also polytypic. Fluorescence *in situ* hybridization showed no abnormality of chromosome 3. Immunoglobulin M was increased to 10.86 g/l (normal range 0.5–1.9) with no paraprotein band being detected on immunofixation. The patient smoked 10 cigarettes per day.

Persistent polyclonal B lymphocytosis is an uncommon condition that is strongly linked to cigarette smoking. The majority of reported cases have been in women. It is linked to HLA-DR7 and familial cases have been reported. Curiously, although the lymphocytes are polyclonal, recurrent cytogenetic abnormalities have been described, particularly i(3)(q), trisomy 3 and dup(3)(q26q29). The presence of binucleated lymphocytes is the distinctive feature that permits the diagnosis to be suspected with a high degree of reliability, even sometimes in patients who lack an increase in the absolute lymphocyte count.

Original publication: Deplano S, Nadal-Melsió E and Bain BJ (2014) Persistent polyclonal B lymphocytosis. *Am J Hematol*, **89**, 224.

Hematology: 101 Morphology Updates, First Edition. Barbara J. Bain.
© 2023 John Wiley & Sons Ltd. Published 2023 by John Wiley & Sons Ltd.

77 Non-hemopoietic cells in the blood and bone marrow

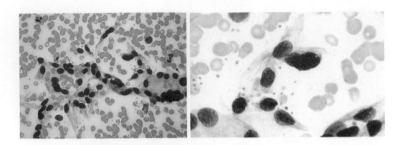

In interpreting peripheral blood and bone marrow films it is necessary to recognize any non-hemopoietic cells that are present. In the blood these can include epithelial cells (either nucleated or not), endothelial cells and even subcutaneous fat cells.[1] Noting the presence and recognizing the nature of such cells is important, both because they may otherwise be misinterpreted as cells of pathological significance and because, in the case of fat cells, they may be sufficiently numerous to interfere with an automated count.[2] The images above show endothelial cells at low power (left) and high power (right). These cells tend to occur in loose sheets and are pleomorphic with round to oval nuclei and variably condensed chromatin. Nuclei may be irregular or grooved and some cells appear to be multinucleated.

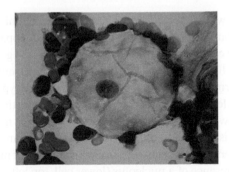

The image above shows an isolated fat cell in the bone marrow. Fat cells or adipocytes are easily recognized in bone marrow fragments. There is less awareness of the morphology of detached single fat cells. This image is of a typical fat cell, showing voluminous weakly staining cytoplasm and an oval nucleus with a delicate chromatin pattern and ill-defined nucleoli. Fat cells have sometimes been confused with storage cells or tumor cells. Recognition of the typical appearance of a fat cell that has become detached from a fragment can avoid confusion.

Original publications: Bain BJ (2013) Endothelial cells. *Am J Hematol*, **88**, 517;Bain BJ (2011) A fat cell in a bone marrow aspirate. *Am J Hematol*, **86**, 66.

References

1 Bain BJ (2022) *Blood Cells: a Practical Guide*, 6th edn. Wiley Blackwell, Oxford, pp. 152–155.
2 Whiteway A and Bain BJ (1999) Artefactual elevation of an automated white cell count following femoral vein puncture. *Clin Lab Haematol*, **21**, 65–68.

Hematology: 101 Morphology Updates, First Edition. Barbara J. Bain.
© 2023 John Wiley & Sons Ltd. Published 2023 by John Wiley & Sons Ltd.

78   It's a black day – metastatic melanoma in the bone marrow

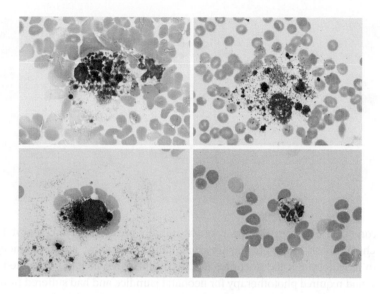

When metastatic cells are present in the bone marrow it is rarely possible to identify the cellular origin of a tumor from morphology alone. An exception is neuroblastoma when neurofibrils are detected.[1] It can also be possible to suspect small cell carcinoma of the lung when the characteristic molding of nuclei is noted, and adenocarcinoma in general when secretory globules are present within tumor cells. One tumor that can be identified with certainty is malignant melanoma. These images are from a 33-year-old man who presented with right axillary lymphadenopathy. He had had a mole on his right shoulder from birth, which had changed characteristics in the preceding year.

A bone marrow aspirate showed abundant macrophages containing melanin (top images). In addition, there were melanoma cells (bottom left), which were sometimes difficult to distinguish from macrophages. Neutrophils in the bone marrow contained melanin (bottom right). The patient's trephine biopsy specimen was black on macroscopic examination. A buffy coat preparation from the peripheral blood of this patient has previously been reported; both circulating melanoma cells and neutrophils containing melanin were detected.[2]

A diagnosis of metastatic melanoma can be confirmed by immunohistochemistry but in this case the clinical and morphological features were sufficient for an unequivocal diagnosis.

Original publication: Bain BJ and Luckit J (2019) It's a black day – metastatic melanoma in the bone marrow. *Am J Hematol*, **94**, 1288–1289.

References

1 Renaudon-Smith E, Arca M, Osei-Yeboah A, Karnick L and Bain BJ (2016) Neuroblastoma in the bone marrow. *Am J Hematol*, **91**, 1272.
2 Swirsky D and Luckit J (1999) Images in haematology: the peripheral blood in metastatic melanoma. *Br J Haematol*, **107**, 219.

Hematology: 101 Morphology Updates, First Edition. Barbara J. Bain.
© 2023 John Wiley & Sons Ltd. Published 2023 by John Wiley & Sons Ltd.

79   Dehydrated hereditary stomatocytosis

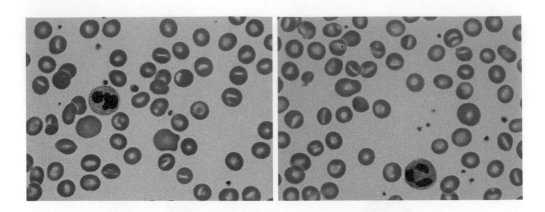

A 27-year-old woman of Northern European origin initially presented at the age of 11 years with symptoms of cholelithiasis. Prior to laparoscopic cholecystectomy, mild normocytic anemia (Hb 100 g/l) with reticulocytosis and unconjugated hyperbilirubinemia had been noted. Her mother reported that she had required phototherapy for neonatal jaundice and had suffered from fatigue and jaundice intermittently since early childhood. There was no antecedent family history of anemia or hemolysis. Extensive investigations for congenital hemolytic anemia failed to identify a clear cause but revealed increased resistance to osmotic lysis (mean cell fragility 3.15 g/l NaCl; reference range 4.0–4.45) with an elevated MCHC (356–378 g/l) and reduced 2,3-DPG content (10.4 µmol/g Hb; reference range 12.7–17.9) and 2,3-DPG/ATP ratio (1.8; reference range 2.4–3.6). Based on this and the red cell morphology, a diagnosis of dehydrated hereditary stomatocytosis (also known as hereditary xerocytosis) was suspected. The blood film showed stomatocytosis, polychromatic macrocytes and occasional target cells and irregularly contracted cells (images). In both dehydrated and overhydrated forms of hereditary stomatocytosis red cells typically exhibit reduced intracellular potassium $[K]_i$ and increased sodium $[Na]_i$ content due to passive cation flux. Measurement of $[Na]_i$ and $[K]_i$ of the patient's red cells was unremarkable but isotopic flux studies using ^{86}Rb as a surrogate demonstrated increased permeability of red cells to potassium with a 4–5-fold increase in K influx resistant to ouabain and bumetanide (to inhibit the Na/K pump and Na/KCl co-transport). Taken together, these findings are consistent with dehydrated hereditary stomatocytosis.

In addition to cholelithiasis, dehydrated hereditary stomatocytosis is associated with several extra-hematological features including perinatal edema, neonatal hepatitis, pseudohyperkalemia and a propensity to non-transfusional iron overload. Splenectomy is contraindicated due to an increased risk of thromboembolism. Inheritance is autosomal dominant. Gain-of-function mutations in *PIEZO*, encoding a mechanosensitive cation channel, and deleterious mutations of *KCNN4*, encoding the Gardos channel, have been implicated. Despite well-compensated hemolysis with an Hb in the range 95–115 g/l the patient suffered significant fatigue and physical limitation. This may be explained by the effect of a low red cell 2,3-DPG concentration on tissue oxygen delivery consistent with which the oxygen dissociation curve was left-shifted with a *p*50 of 20.76 mmHg (reference range 27–33).

With thanks to Professor Gordon Stewart for the intracellular cation flux studies.

Original publication: Layton DM and Bain BJ (2016) Dehydrated hereditary stomatocytosis. *Am J Hematol*, **91**, 266.

80 Circulating lymphoma cells in intravascular large B-cell lymphoma

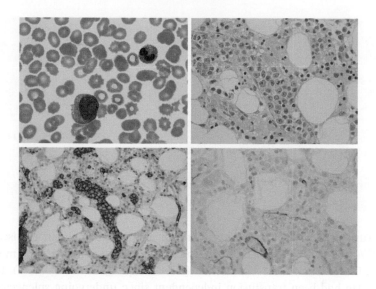

A 92-year-old Caucasian man who was dialysis dependent due to autosomal dominant polycystic kidney disease presented with thrombocytopenia and weight loss. On examination, his spleen was palpable 6 cm below the left costal margin but there was no palpable lymphadenopathy. His blood count showed a stable Hb of 101 g/l but a markedly reduced platelet count of 44×10^9/l and WBC of 19.3×10^9/l. An automated differential count suggested that there was monocytosis and lympho-cytosis but a blood film showed that the abnormal cells were all lymphoid, with the count being neutrophils 3.8×10^9/l, lymphoid cells 12×10^9/l, monocytes 0.6×10^9/l and eosinophils 0.1×10^9/l. The abnormal cells were medium to large with a high nucleocytoplasmic ratio, a single large but indistinct nucleolus and moderately basophilic cytoplasm (top left image). Flow cytometric immunophenotyping on peripheral blood showed the abnormal cells to be large B cells expressing CD19, CD20, CD79b and BCL2 with κ light chain restriction. They were negative for CD5, CD10, CD23, CD43 and CD200. The proliferation fraction (Ki-67 positive) was 67%. A bone marrow aspi-rate was hemodilute, with some maturing hematopoietic cells but also lymphoid cells similar to those seen in the peripheral blood. Trephine biopsy sections were hypercellular for the patient's advanced age and showed atypical medium to large cells, with prominent nucleoli, within sinusoids and capillaries (top right, H&E). Immunohistochemistry for CD20 (bottom left) confirmed that the abnormal cells were largely intravascular with smaller numbers of single cells and small aggregates in the interstitium. The neoplastic cells also expressed CD19 and CD79a while CD10, MUM1/IRF4 and EBER were negative. CD34 staining of endothelial cells confirmed their intravascular location (bottom right). The patient was given a trial of high dose corticosteroids but died shortly afterwards.

The prognosis of intravascular large B-cell lymphoma is usually adverse, at least in part due to the difficulty in making the diagnosis. In our patient the unusual presence of circulating cells led to examination of the bone marrow. Despite the speedy diagnosis, treatment in this elderly dialysis-dependent man was unsuccessful.

Original publication: Fordham NJ, O'Connor S, Stern S, Nikolova V and Bain BJ (2017) Circulating lymphoma cells in intravascular large B-cell lymphoma. *Am J Hematol*, **92**, 311.

Hematology: 101 Morphology Updates, First Edition. Barbara J. Bain.
© 2023 John Wiley & Sons Ltd. Published 2023 by John Wiley & Sons Ltd.

81 Unusual inclusions in hemoglobin H disease post-splenectomy

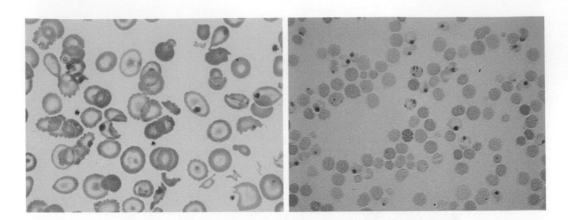

A 62-year-old Turkish Cypriot man who was known to have hemoglobin H disease with the genotype $--^{MED}/\alpha^T\alpha$ due to coinheritance of α^0 thalassemia and a non-deletional mutation of the $\alpha2$ globin gene, IVS-I donor site 5 bp deletion (HBA2:c.95+2_95+6delTGAGG), that prevents mRNA splicing. He had been transfusion independent since undergoing splenectomy 20 years earlier. Other past medical history included recurrent gout, pulmonary embolism, stage IV chronic kidney disease, chronic hepatitis B infection and asthma. On routine tests his Hb had dropped to 67 g/l (from a baseline of around 90–100 g/l) with an MCV of 56 fl (baseline 67 fl). He had mild shortness of breath. This deterioration was found to be due to superimposed iron deficiency related to chronic gastrointestinal bleeding from colonic angiodysplastic lesions found on colonoscopy.

The blood film showed multiple abnormalities (left image). There was severe microcytosis, hypochromia, striking poikilocytosis and target cells. Post-splenectomy changes were also seen: acanthocytes, Howell–Jolly bodies and giant platelets. There were red cell fragments, some of which were acanthocytic while others contained Howell–Jolly bodies or Pappenheimer bodies. Of particular interest, however, were unusual inclusions that are specific for hemoglobin H disease post-splenectomy. These were round and dense with the staining characteristics of hemoglobin. They appeared to be bound to the red cell membrane. They resembled the α chain inclusions that can be seen in β thalassemia major or intermedia but appeared denser. A hemoglobin H preparation was positive and showed Heinz bodies in approaching one-third of cells (right).

Red cell inclusions such as these, representing precipitated β globin tetramers, are not often observed in hemoglobin H disease since, with intact splenic function, they are removed by the spleen. Following splenectomy, they are present in peripheral blood erythrocytes *in vivo* and are readily observed on a standard May–Grünwald–Giemsa stain. Post-splenectomy two types of inclusion can thus be observed in a hemoglobin H preparation: (i) preformed Heinz bodies that are present *in vivo*; and (ii) characteristic 'golf-ball' cells resulting from *in vitro* precipitation of dissolved hemoglobin H.

Original publication: Spencer-Chapman M, Luqmani A, Layton DM and Bain BJ (2018) Unusual inclusions in hemoglobin H disease post-splenectomy. *Am J Hematol*, **93**, 963–964.

82 An unexpectedly bizarre blood film in hemoglobin H disease

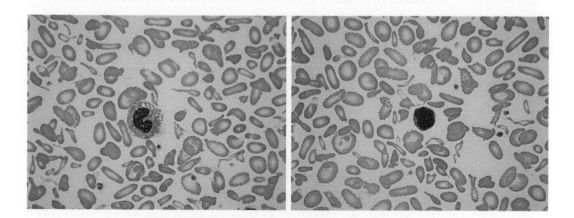

A 3-year-old boy from Saudi Arabia had been followed up with a diagnosis of hemoglobin H disease, identified in a neonatal screening program. His parents were second cousins. Alpha globin genotyping at birth had revealed homozygosity for the non-deletional Saudi-T mutation ($\alpha^T\alpha/\alpha^T\alpha$), both parents being carriers. He had remained clinically asymptomatic. Red cell indices at the age of 3 years were RBC 3.84×10^{12}/l, Hb 80 g/l, MCV 61 fl, MCH 20.9 pg, MCHC 34 g/l, red cell distribution width 34.8 (normal range 11.8–14.3) and reticulocyte count 447×10^9/l. His blood film, in addition to microcytosis, showed gross poikilocytosis with the abnormal red cells including elliptocytes and many bizarre poikilocytes, some very small (images). Although poikilocytosis can be prominent in hemoglobin H disease, we had not seen a case with this degree of abnormality. The film was reminiscent of hereditary pyropoikilocytosis. For this reason red cell membrane studies were initiated. An eosin-5′-maleimide (EMA) binding assay[1] gave a twin peak fluorescence profile: 57% of phenotypically normal red cells (MCF = 64.7 units) and 42% having reduced fluorescence (MCF = 32.4 units). Spectrin (Sp) analysis showed a spectrin dimer content of 25.6%, compared with 7.4 % for a normal control within the same assay. In limited trypsin digestion of spectrin, Sp$\alpha^{I/50}$ variant was detected in the child and absent in the mother. Neither had an increase of Sp$\alpha^{V/41}$ tryptic peptide, thus excluding the presence of the polymorphic low expression allele Spα^{LELY}. Maternal red cells were not elliptocytic and had normal spectrin dimer content and normal EMA binding. A paternal sample was not available for testing.

Although the EMA-binding assay has high specificity and sensitivity for hereditary spherocytosis, it can also be abnormal in hereditary pyropoikilocytosis, South-East Asian ovalocytosis, cryohydrocytosis and some cases of type II congenital dyserythropoietic anemia.[2] Hereditary elliptocytosis usually gives a single fluorescence peak within the range for normal red cells but occasionally twin peaks are observed. In this child with coinheritance of hemoglobin H disease and hereditary elliptocytosis, the severity of the hemolytic anemia is that expected in uncomplicated hemoglobin H disease but the bizarre blood film reflects an interaction of the hemoglobinopathy and the membrane defect.

Original publication: Chakravorty S, King MJ and Bain BJ (2012) An unexpectedly bizarre blood film in hemoglobin H disease. *Am J Hematol*, **87**, 1104.

Hematology: 101 Morphology Updates, First Edition. Barbara J. Bain.
© 2023 John Wiley & Sons Ltd. Published 2023 by John Wiley & Sons Ltd.

References

1 King MJ, Telfer P, MacKinnon H, Langabeer L, McMahon C, Darbyshire P and Dhermy D (2008) Using the eosin-5-maleimide binding test in the differential diagnosis of hereditary spherocytosis and hereditary pyropoikilocytosis. *Cytometry B Clin Cytom*, **74**, 244–250.
2 King MJ, Behrens J, Rogers C, Flynn C, Greenwood D and Chambers K (2000) Rapid flow cytometric test for the diagnosis of membrane cytoskeleton-associated haemolytic anaemia. *Br J Haematol*, **111**, 924–933.

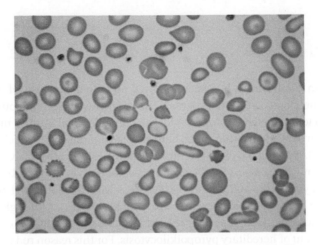

Blood film of an other patient with typical hemoglobin H disease showing anisocytosis, poikilocytosis including a teardrop poikilocyte and several elliptocytes, hypochromia, microcytosis and polychromatic cells.

83 Acute myeloid leukemia with a severe coagulopathy and t(8;16)(p11;p13)

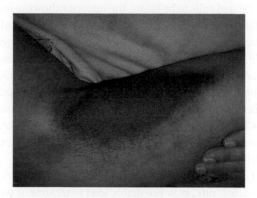

A 43-year-old man presented to the emergency department with a history of hematuria and heavy bruising (image above shows axillary bruising). Blood tests showed WBC 32.4×10^9/l, Hb 158 g/l and platelet count of 97×10^9/l. Coagulation studies showed fibrinogen <0.40 g/l (normal range 1.9–4.3), prothrombin time (PT) 33.6 seconds (12.8–17.4), activated partial thromboplastin time (APTT) 54.6 seconds (25–35) and D-dimer >20,000 µg/l (<500).

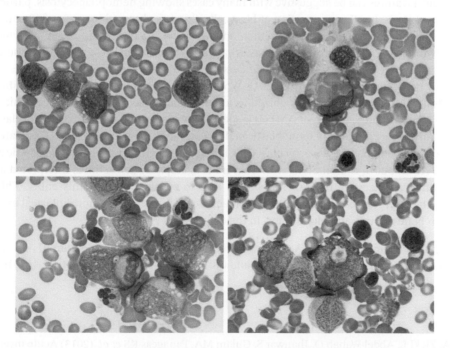

Urgent assessment of the blood film revealed a mixture of granular myeloblasts and a smaller proportion of monoblastoid forms; occasional cells had a bilobed nucleus (top left). The presence of disseminated intravascular coagulation (DIC) in a patient with acute myeloid leukemia (AML) raised the possibility of acute promyelocytic leukemia (APL). It being Friday evening, molecular exclusion of a *PML::RARA* would not be possible for 72 hours, so a decision was made to commence

Hematology: 101 Morphology Updates, First Edition. Barbara J. Bain.
© 2023 John Wiley & Sons Ltd. Published 2023 by John Wiley & Sons Ltd.

all-*trans*-retinoic acid (ATRA) and dexamethasone together with cryoprecipitate, fibrinogen concentrate and fresh frozen plasma. There was no improvement in the severe hypofibrinogenemia, and the WBC increased to 53.9×10^9 over 3 days. A bone marrow aspiration was performed with the aspirate showing 75% of nucleated cells to be blasts, the majority showing monoblastoid morphology. Notably, distinctive examples of erythrophagocytosis by blast cells were also seen, with occasional phagocytosis also of neutrophils and platelets (top right and bottom images). Immunophenotyping showed early monocytic differentiation with expression of myeloperoxidase, CD33 (strong), CD15, CD36, CD64 (weak) and HLA-DR. There was no expression of CD11b, CD13, CD14, CD34 or CD300e. The bone marrow morphology, in combination with the clinical features, led to the suspicion of AML with t(8;16); this was confirmed shortly afterwards when G-banded chromosome analysis showed 46,XY,t(8;16)(p11;p13)[7]/47,idem,+8,i(8)(q10)[3]. Treatment with ATRA was discontinued and combination chemotherapy with daunorubicin, cytarabine and gemtuzumab ozogamicin was initiated, with prompt reversal of the hypofibrinogenemia and leukocytosis.

An acute presentation of AML with a severe coagulopathy is characteristic of APL and, because of the still significant early death rate, may justifiably lead to treatment with ATRA prior to confirmation or exclusion of the diagnosis. However, it is important to note that the rare entity of AML with t(8;16) and a *KAT6A::CREBBP* fusion gene enters into the differential diagnosis. The frequent lack of expression of CD34 is such cases (14 of 17 negative in one series of patients[1] and 14 of 15 in a second[2]) may strengthen the suspicion of APL. Other stem cell markers, CD117 and CD133, are also often negative. However, in contrast to APL, HLA-DR is generally positive. The bone marrow morphological features can be suggestive with many cases showing hemophagocytosis, particularly erythrophagocytosis, by a low proportion of blast cells; this was observed in seven of 13 patients in one series of patients,[3] in eight of 11 in another[1] and in seven of 15 in a third.[2] Although cytochemistry is now rarely used in leukemia diagnosis, it can be informative in these cases since there is an otherwise unusual coexpression of non-specific esterase and quite strong myeloperoxidase.

Although not a distinct entity in the 2016 World Health Organization (WHO) classification, AML with t(8;16) is increasingly recognized as an alternate cause of hypofibrinogenemia and a bleeding diathesis at presentation of AML. It is specifically recognized in the 2022 WHO classification. The morphology is typically monoblastic or, less commonly, myelomonocytic. Cases are often therapy related. Hemophagocytosis by blast cells is associated with a variety of cytogenetic abnormalities in AML, but in combination with a severe coagulopathy (sometimes interpreted as DIC and sometimes as excessive fibrinolysis) should lead to consideration of this entity. This case underlines the importance of an integrated approach, including a careful morphological examination of both bone marrow and blood films, in the diagnosis of AML.

Original publication: Hastings A, Apperley JF, Nadal-Melsio E, Brown L and Bain BJ (2021) Acute myeloid leukemia with a severe coagulopathy and t(8;16)(p11;p13). *Am J Hematol*, **96**, 163–164.

References

1 Diab A, Zickl L, Abdel-Wahab O, Jhanwar S, Gulam MA, Panageas KS *et al.* (2013) Acute myeloid leukemia with translocation t(8;16) presents with features which mimic acute promyelocytic leukemia and is associated with poor prognosis. *Leuk Res*, **37**, 32–36.

2 Xie W, Hu S, Xu J, Chen Z, Chen Z, Medeiros LJ and Tang G (2019) Acute myeloid leukemia with t(8;16)(p11.2;p13.3)/*KAT6A-CREBBP* in adults. *Ann Hematol*, **98**, 1149–1157.

3 Haferlach T, Kohlmann A, Klein HU, Ruckert C, Dugas M, Williams PM *et al.* (2009) AML with translocation t(8;16)(p11;p13) demonstrates unique cytomorphological, cytogenetic, molecular and prognostic features. *Leukemia*, **23**, 934–943.

84 Cold autoimmune hemolytic anemia secondary to atypical pneumonia

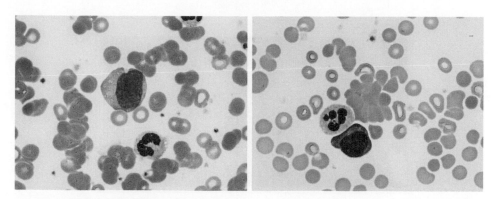

A 44-year-old Caucasian woman presented with a 10-day history of dry cough, exertional dyspnea, myalgia and fever. Symptoms persisted despite having completed a course of amoxicillin and erythromycin. She was an ex-smoker of 15 pack years living in rented accommodation with mold and with bird exposure, having a canary as a pet. On presentation, she was hemodynamically stable and afebrile with a normal respiratory rate, an oxygen saturation of 96% on air and sinus tachycardia of 98 beats per minute. There were fine crackles throughout the lung fields and right upper zone and reduced air entry with increased vocal resonance. The spleen tip was palpable. Chest radiography showed a focal area of patchy consolidation in the right upper lobe.

Her blood count on admission showed a raised WBC of 16×10^9/l and platelet count of 494×10^9/l. She was anemic with an Hb of 84 g/l, MCV of 77 fl and reticulocyte count of 176×10^9/l. Renal and liver function (including total bilirubin) were normal, with a raised C-reactive protein of 34 mg/l (normal range <6) and a high lactate dehydrogenase of 639 iu/l (125–243). An automated differential count suggested monocytosis but a manual differential count showed neutrophils 8.3×10^9/l, lymphocytes 6.1×10^9/l and monocytes 0.6×10^9/l. Her blood film showed red cell agglutinates, polychromatic macrocytes, nucleated red blood cells and atypical lymphocytes. Some of the lymphocytes were large with increased cytoplasmic basophilia (left image). Others had plasmacytoid features with an eccentric nucleus and a Golgi zone (right). There were occasional examples of erythrophagocytosis. A direct antiglobulin test was positive for C3d.

On the basis of the clinical and the hematological findings, cold antibody-induced hemolytic anemia secondary to *Mycoplasma pneumoniae* infection was suspected. This was confirmed by demonstration of both IgG and IgM antibodies, indicative of a recent infection. PCR for Epstein–Barr virus (EBV) was negative, as were laboratory tests for multiple other organisms. A computed tomography scan showed 'tree in bud' airspace consolidation and associated mediastinal lymphadenopathy. The patient made a rapid recovery with intravenous co-amoxiclav and oral clarithromycin. Subsequent imaging showed complete resolution of radiographic abnormalities.

The combination of cold autoimmune hemolytic anemia and atypical lymphocytes suggests either EBV or mycoplasma infection. As in this patient, consideration of the clinical as well as hematological features can indicate the correct diagnosis.

Original publication: Atta M, Brannigan ET and Bain BJ (2017) Cold autoimmune hemolytic anemia secondary to atypical pneumonia. *Am J Hematol*, **92**, 109.

85 A confusing 'white cell count' – circulating micromegakaryocytes in post-thrombocythemia myelofibrosis

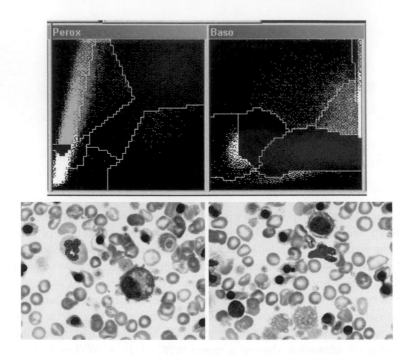

A 55-year-old woman presented to the emergency department with rectal bleeding. She had a past medical history of *JAK2* V617F-positive essential thrombocythemia transforming to myelofibrosis with hypersplenism; for this she had had a splenectomy. A blood count on a Siemens ADVIA 2120i showed WBC 223.8 × 10⁹/l (neutrophils 12.9%, lymphocytes 18.7%, monocytes 4.6%, eosinophils 0.2%, basophils 1.7% and 'large unstained cells' (peroxidase-negative cells) 63.6%), Hb 81 g/l, platelet count 494 × 10⁹/l and nucleated red blood cells (NRBC) 256/100 white cells. In view of the very abnormal blood count and scatter plots (top image), microscopic examination was performed. This confirmed the presence of numerous NRBC, these being 75% of the total nucleated cell count (TNCC). However, in addition there were numerous micromegakaryocytes and apparently bare megakaryocyte nuclei (bottom images), constituting 15% of the TNCC. The true WBC was 22.38 × 10⁹/l. Furthermore, microscopy highlighted the presence of numerous platelets including giant platelets; some were hypogranular and some were aggregated. The true platelet count was thus higher than that provided by the analyzer.

Micromegakaryocytes are a feature of myelodysplastic syndromes and myeloproliferative/myelodysplastic neoplasms. However, they can also occur during evolution of myeloproliferative neoplasms. In the absence of a spleen, considerable numbers can appear in the peripheral blood and being of similar size to white cells can be counted as such by automated instruments.

Original publication: Sale S, Rocco V, Pancione Y, Carri DD, Fumi M and Bain BJ (2019) A confusing "white cell count": circulating micromegakaryocytes in post-thrombocythemia myelofibrosis. *Am J Hematol*, **94**, 617–618.

86 Diagnosis of follicular lymphoma from the peripheral blood

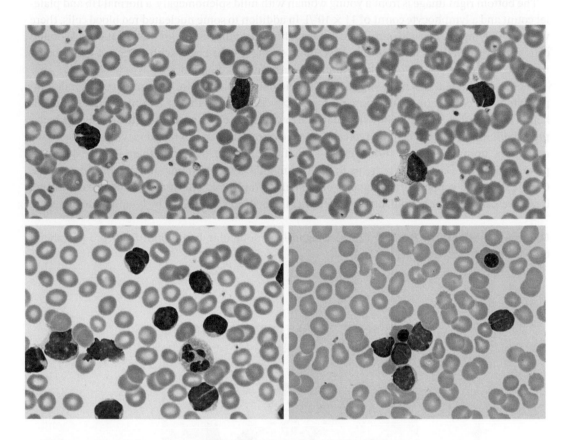

In patients with stage IV disease, it is often possible to diagnose follicular lymphoma from the peripheral blood, with the blood film being crucial for indicating this diagnosis. The top two images are from a middle-aged man with generalized lymphadenopathy and a previous history of splenectomy following a road traffic accident. His Hb and platelet count were normal but the lymphocyte count was 10.5×10^9/l. His blood film showed the expected post-splenectomy changes including an increase of cytologically normal lymphocytes, among which large granular lymphocytes were prominent. However, there was also a population of small lymphocytes with dense chromatin and deep narrow nuclear clefts (top images) suggesting the presence of follicular lymphoma in addition to post-splenectomy lymphocytosis. Immunophenotyping supported these diagnoses, showing 55% CD8+ T cells, 40% CD4+ T cells and 17% clonal B cells expressing CD19, CD79b, CD10 and strong λ light chain and not expressing CD5. The diagnosis of follicular lymphoma was confirmed by lymph node biopsy.

The bottom left image is from another middle-aged man with lymphadenopathy, mild anemia (Hb 128 g/l), a normal platelet count and a lymphocyte count of 89×10^9/l. His blood film showed small and medium-sized lymphocytes with dense chromatin; some were nucleolated and some had deep narrow nuclear clefts. Immunophenotyping showed clonal B cells expressing FMC7 and strong κ light chain and not expressing CD5 or CD10; follicular lymphoma was confirmed on a lymph node biopsy.

The bottom right image is from a young woman with mild splenomegaly, a normal Hb and platelet count and a lymphocyte count of 11×10^9/l. In addition to some nucleated red blood cells, there were small lymphocytes with similar features to those described in the first two patients. Clonal B cells expressed CD19, CD10, CD79b, FMC7 and strong κ light chain with CD5 and CD23 being negative. The diagnosis of follicular lymphoma was confirmed on biopsy.

The blood film features most suggestive of follicular lymphoma are the presence of small lymphocytes with chromatin that is more uniformly condensed than that of chronic lymphocytic leukemia cells, with some cells showing deep, narrow nuclear clefts and with smear cells not being prominent. Immunophenotyping typically shows expression of pan-B markers with strong surface membrane immunoglobulin and no expression of CD23 or CD200. When there is some degree of pleomorphism, CD5 negativity helps to exclude mantle cell lymphoma and, when it is positive, CD10 expression supports the diagnosis of follicular lymphoma. This diagnosis can be confirmed by fluorescence *in situ* hybridization on peripheral blood cells, as well as by lymph node biopsy.

Original publication: Bain BJ (2018) Diagnosis of follicular lymphoma from the peripheral blood. *Am J Hematol*, **93**, 1111–1112.

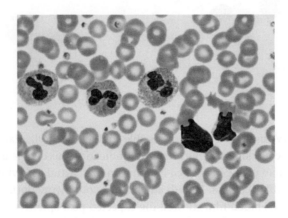

A further image of the blood film of the first patient showing two typical follicular lymphoma cells.

87 Transformation of follicular lymphoma

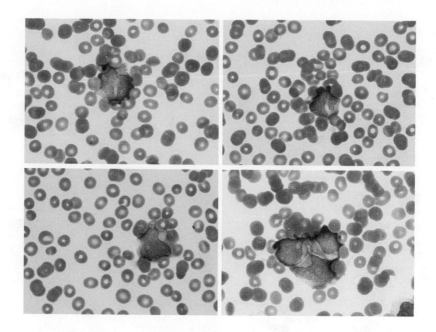

A 73-year-old woman presented with a 3-week history of fevers and sudden weight loss. Her past medical history included stage IV follicular lymphoma diagnosed 9 years earlier, which had been treated with rituximab-based immunochemotherapy. Clinical examination showed bilateral cervical and axillary lymphadenopathy and splenomegaly.

She was found to be bicytopenic with Hb 99 g/l, WBC 7.4×10^9/l and platelet count 8×10^9/l. The lymphocyte count was 5.2×10^9/l and neutrophil count 1.5×10^9/l. A blood film showed a minor population of small lymphocytes with cleft nuclei, consistent with the previous diagnosis of follicular lymphoma. However, in addition, there were large lymphoid cells with prominent nucleoli, some with a high nucleocytoplasmic ratio. Some had cleft nuclei and were consistent with large centrocytes (top images). Others had round or oval nuclei and were consistent with centroblasts (bottom images). Bone marrow fluorescence *in situ* hybridization showed 36 out of 40 cells to have an *IGH::BCL2* rearrangement, as expected in follicular lymphoma. In addition, 24 of 40 cells showed a *MYC* rearrangement. These findings were indicative of a high grade transformation of follicular lymphoma, with *MYC* rearrangement and with *BCL2* juxtaposed to the *IGH* locus.

The peripheral blood film can thus reveal not only a diagnosis of follicular lymphoma[1] but also of high grade transformation of this lymphoma.

Original publication: Siow W, Burton K, Eagleton H, Campbell C and Bain BJ (2018) Transformation of follicular lymphoma. *Am J Hematol*, **93**, 1292–1293.

Reference

1 Bain BJ. Diagnosis of follicular lymphoma from the peripheral blood. *Am J Hematol*, **93**, 1111–1112.

Hematology: 101 Morphology Updates, First Edition. Barbara J. Bain.
© 2023 John Wiley & Sons Ltd. Published 2023 by John Wiley & Sons Ltd.

88 Cytology of systemic mastocytosis

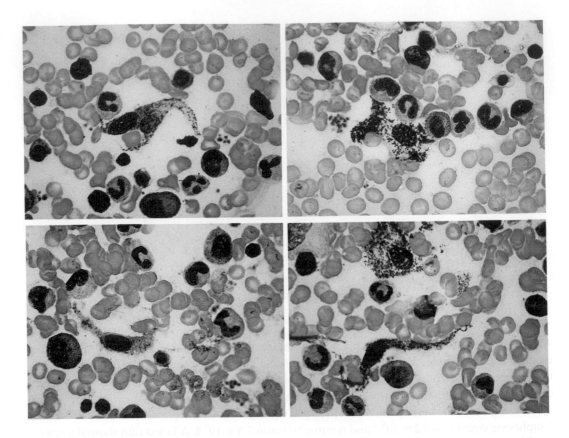

A 65-year-old woman with a 15-year history of urticaria pigmentosa on her trunk, arms and legs was referred for further investigation when she was found to have elevation of serum tryptase to 190 μg/l (normal range 2–14). She had noticed worsening of her skin lesions but was otherwise in good health. A blood count, blood film and liver and renal function tests were normal. A bone marrow aspirate showed significant numbers of cytologically abnormal mast cells. Many of these had long and elegant tails, sometimes even three per cell (top left image). Some of the spindle-shaped cells (bottom left) and other mast cells (right) were hypogranular. A morphological diagnosis of systemic mastocytosis was confirmed by demonstration of a *KIT* D816V mutation in bone marrow cells.

Neoplastic mast cells may be undetectable in bone marrow aspirates from patients with systemic mastocytosis, even when trephine biopsy sections show significant infiltration. When present, they are usually cytologically abnormal so that the diagnosis can be made from marrow films. Hypogranular and spindle-shaped cells are characteristic. In addition, there may be aberrant expression of CD2 and CD25 and when mast cells are present in significant numbers in the aspirate a *KIT* mutation, usually *KIT* D816V, is likely to be detected.

Original publication: Bain BJ and Marks AJ (2009) Cytology of systemic mastocytosis. *Am J Hematol*, **84**, 842.

Hematology: 101 Morphology Updates, First Edition. Barbara J. Bain.
© 2023 John Wiley & Sons Ltd. Published 2023 by John Wiley & Sons Ltd.

89 Systemic mastocytosis – the importance of looking within bone marrow fragments

 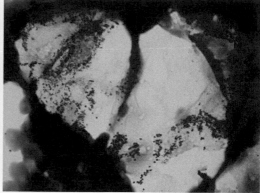

The detection of neoplastic mast cells in a bone marrow aspirate can be difficult, even when trephine biopsy sections show an abnormal infiltrate. This is the result of infiltrates being focal, often in a paratrabecular or perivascular site, with reticulin fibrosis hindering their aspiration. In addition, mast cells can remain trapped within bone marrow particles rather than spilling into the cellular trails behind them. When mastocytosis is suspected it is important to look at the trails immediately behind the particles where a few cytologically abnormal mast cells may be found. In addition, it is also important to look within the fragments, which may be found to harbour numerous mast cells, recognizable from their brightly staining granules. Some cytological abnormalities may be apparent, even in the cells within fragments. The left image shows the typical spindle shape of a neoplastic cell, which has three cytoplasmic tails. The right image shows hypogranularity, also a typical feature of neoplastic mast cells.

Another useful technique is to make a squash preparation of a bone marrow particle. The International Council for Standardisation in Haematology guidelines for bone marrow examination recommend that this be done, in addition to wedge-spread films, on all bone marrow aspirates. This is good practice and will facilitate the diagnosis not only of systemic mastocytosis but also of some cases of multiple myeloma.

Finally, a trephine biopsy should be performed whenever systemic mastocytosis is suspected.

Original publication: Bain BJ, Khoder A, Milojkovic D and Bernard T (2014) Systemic mastocytosis – the importance of looking within bone marrow fragments. *Am J Hematol*, **89**, 109–110.

90 Schistocytosis is not always microangiopathic hemolytic anemia

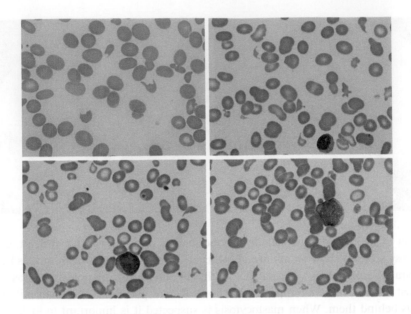

A 75-year-old woman presented elsewhere with gradually increasing shortness of breath, weight loss and anorexia. Her blood count showed anemia and mild thrombocytopenia and a blood film showed red cell fragments (schistocytes). Thrombotic thrombocytopenic purpura (TTP) was suspected and the patient was referred for plasma exchange. On arrival, the blood count and blood film examination were repeated. The blood count showed WBC 6.8×10^9/l, Hb 80 g/l, MCV 91.6 fl, red cell distribution width 26.8%, platelet count 111×10^9/l and reticulocyte count 49.7×10^9/l. In addition to frequent schistocytes (top images), the blood film showed macrocytes, elliptocytes, Howell–Jolly bodies, occasional nucleated red cells, platelet anisocytosis and occasional giant platelets. Furthermore there were 4% blast cells (lower images). Lactate dehydrogenase was elevated to 666 iu/l. Renal and liver function tests, a coagulation screen, C-reactive protein and hematinic assays were all normal. The provisional diagnosis was doubted and bone marrow aspiration was performed. This showed trilineage dysplasia with 7% blast cells with occasional blast cells containing Auer rods. Cytogenetic analysis showed a complex karyotype with monosomy of chromosomes 5, 7 and 17 (*TP53* deletion being demonstrated on fluorescence *in situ* hybridization). The diagnosis was revised to myelodysplastic syndrome with excess blasts 2. Subsequently, the ADAMTS13 level was found to be normal.

Rapid diagnosis and treatment of TTP is important and the index of suspicion must be high. However, it is important to be aware of other causes of schistocytosis. The dyserythropoiesis of a myelodysplastic syndrome (MDS) can lead to the presence of schistocytes and we are aware of another patient with MDS in whom plasma exchange for suspected TTP was carried out before bone marrow examination revealed the correct diagnosis. Severe megaloblastic anemia can also have marked schistocytosis so that TTP is suspected.

Original publication: Lofaro T and Bain BJ (2020) Schistocytosis is not always microangiopathic hemolytic anemia. *Am J Hematol*, **95**, 1421–1422.

Hematology: 101 Morphology Updates, First Edition. Barbara J. Bain.
© 2023 John Wiley & Sons Ltd. Published 2023 by John Wiley & Sons Ltd.

91   Hemoglobin C disease

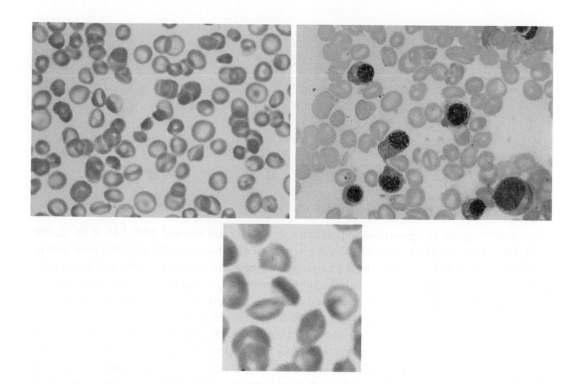

A 56-year-old Afro-Caribbean man presented with gall stones. His blood count showed RBC 4.59×10^{12}/l, Hb 135 g/l, Hct 0.41 l/l, MCV 89.6 fl, MCH 29.4 pg and MCHC 328 g/l. His blood film showed well-hemoglobinized cells with moderate numbers of irregularly contracted cells and target cells (top left image). There were small numbers of cells with the hemoglobin retracted to one or both sides of a cell, and small numbers of cells containing hemoglobin C crystals (bottom image). Hemoglobin electrophoresis showed hemoglobin C plus hemoglobin A_2 only. A diagnosis of hemoglobin C disease was made. The coexistence of irregularly contracted cells and target cells is very characteristic of this condition. Hemoglobin C crystals are much less often observed.

The patient consented to bone marrow aspiration during cholecystectomy. The aspirate showed erythroid hyperplasia and dyserythropoiesis. The top right image shows two cytologically disparate erythroblasts joined by a cytoplasmic bridge. A number of the erythroblasts show an irregular nuclear margin and an abnormal chromatin pattern, characteristic features of hemoglobin C homozygosity, which correlate with the disorganized nuclei with intranuclear clefts and loss and irregularity of parts of the nuclear membrane that have been described on ultrastructural examination.[1]

Original publication: Bain BJ (2015) Hemoglobin C disease. *Am J Hematol*, **90**, 174.

Reference

1 Wickramasinghe SN, Akinyanju OO and Hughes M (1982) Dyserythropoiesis in homozygous haemoglobin C disease. *Clin Lab Haematol*, **4**, 373–381.

Hematology: 101 Morphology Updates, First Edition. Barbara J. Bain.
© 2023 John Wiley & Sons Ltd. Published 2023 by John Wiley & Sons Ltd.

92 Hemoglobin St Mary's

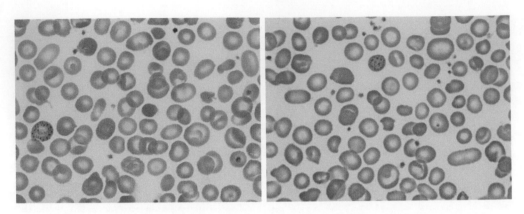

A 5-year-old English girl presented with anemia. A blood count showed RBC 3.85×10^{12}/l, Hb 87 g/l, MCV 81 fl, MCH 22.5 pg, MCHC 278 g/l, red cell distribution width 25.7 and reticulocyte count 443×10^9/l (11.5%). Her blood film showed irregularly contracted cells and basophilic stippling (images). There were also polychromatic macrocytes, occasional cells with hemoglobin retracted to one side of the cell and occasional cells with siderotic granules. There was no history of exposure to drugs or other oxidants. Further tests were therefore done.

High performance liquid chromatography (HPLC) on a BioRad Variant instrument showed a peak in the hemoglobin S window (retention time 4.6 minutes) measuring 7.4%. No abnormal band was apparent on cellulose acetate electrophoresis at alkaline pH or agarose gel electrophoresis at acid pH. However, on isoelectric focusing a variant hemoglobin was identified, which separated from hemoglobin A and comprised about half of the total hemoglobin. A heat test for an unstable hemoglobin was positive. Electrospray tandem mass spectrometry identified a novel hemoglobin variant, $\beta101^{(Glu \rightarrow Ala)}$, which we designated hemoglobin St Mary's,[1] subsequently also referred to as hemoglobin Youngstown. Bone marrow examination showed dyserythropoiesis and 30% ring sideroblasts. Her father, who had an Hb of 143 g/l, a reticulocyte count of 93×10^9/l and smaller numbers of irregularly contracted cells, was found to be a carrier of the same variant hemoglobin. The explanation for the low percentage abnormal peak on HPLC is that this represents denatured hemoglobin St Mary's, the intact variant hemoglobin having the same retention time as hemoglobin A.

Irregularly contracted cells suggest a limited range of diagnostic possibilities: defects of the pentose shunt, exposure to endogenous or exogenous oxidants or a hemoglobinopathy (hemoglobin C, hemoglobin E or an unstable hemoglobin). Small numbers can also be seen in β thalassemia heterozygosity. Dyserythropoiesis can be a feature of a hemoglobinopathy.

Original publication: Wild BJ, Phelan LO and Bain BJ (2016) Hemoglobin St Mary's. *Am J Hematol*, **91**, 735.

Reference

1 Wild BJ, Green BN, Lalloz MRA, Erten S, Amos RJ, Horn J *et al.* (1998) Identification of novel haemoglobin variants Hb Hackney $\alpha27^{(Glu \rightarrow Ala)}$, Hb St Mary's, $\beta101^{(Glu \rightarrow Ala)}$ by mass spectrometry. *Br J Haematol*, **101**, 53, S1, Abstract 125.

Hematology: 101 Morphology Updates, First Edition. Barbara J. Bain.

93 Congenital sideroblastic anemia in a female

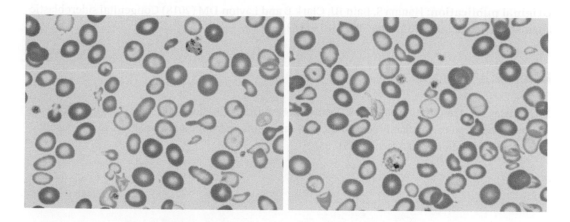

A 20-year-old female of Pakistani descent was referred because of a microcytic hypochromic anemia refractory to oral iron, which had been present since early childhood. A diagnosis of thalassemia intermedia had previously been suspected but high performance liquid chromatography showed normal hemoglobin A_2 and F levels and no mutation of the α or β globin genes was detected by DNA sequencing or deletion analysis. Her parents were consanguineous. Neither parent had any hematological abnormality. On physical examination she was of short stature (<4th centile) but exhibited no dysmorphism or other somatic abnormalities. A blood count showed Hb 70 g/l with normal red cell count, MCV 66.6 fl and MCH 18.3 pg. The reticulocyte count was 58.6×10^9/l and there were no biochemical markers of hemolysis. Transferrin saturation and serum ferritin were raised at 96% and 349 µg/l, respectively. A blood film was dimorphic with a large proportion of poorly hemoglobinized microcytes (images); there was anisopoikilocytosis with dacrocytes, elliptocytes and occasional target cells. Pappenheimer bodies were present, some being very large. The morphological features suggested sideroblastic anemia, in the context most likely congenital. To confirm this, next generation sequencing of a comprehensive gene panel for inherited red cell disorders was undertaken. This identified a homozygous mutation of the *SLC25A38* gene c.401G>A p.(Arg134His). This involves a highly conserved amino acid and has previously been described as a pathogenic variant in autosomal recessive pyridoxine-refractory sideroblastic anemia.[1] Both parents were heterozygous for this variant.

Congenital sideroblastic anemias are a rare group of inherited disorders characterized by the pathological deposition of iron in the mitochondria of erythroid precursors in the bone marrow leading to ineffective erythropoiesis with iron overload and hypochromic red cells in the peripheral blood. They can be divided into syndromic and non-syndromic forms and are genetically heterogeneous with nine monogenic forms and others due to mitochondrial DNA deletion. Non-syndromic forms are either X-linked or autosomal recessive. Their morphological hallmark is the presence of ring sideroblasts in the bone marrow. These represent iron-laden mitochondria which form a ring around the nucleus of erythrocyte precursors. This feature points to a common pathophysiology due to genetic mutations that disrupt heme-iron biosynthesis in the mitochondria. *SLC25A38* encodes an erythroid-specific mitochondrial carrier protein that transports glycine into mitochondria for the first step in heme synthesis. In this case molecular diagnosis prompted by characteristic morphological atypia obviated the need for bone marrow examination and enabled genetic counselling and screening to be provided to the patient and her family.

Original publication: Hanina S, Bain BJ, Clark B and Layton DM (2018) Congenital sideroblastic anemia in a female. *Am J Hematol*, **93**, 1181–1182.

Reference

1 Guernsey DL, Jiang H, Campagna DR, Evans SC, Ferguson M, Kellogg MD *et al.* (2009) Mutations in mitochondrial carrier family gene *SLC25A38* cause nonsyndromic autosomal recessive congenital sideroblastic anemia. *Nat Genet*, **41**, 651–653.

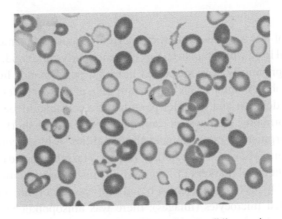

A further image showing dimorphism, poikilocytosis and a cell containing Pappenheimer bodies.

94 A puzzling case of methemoglobinemia

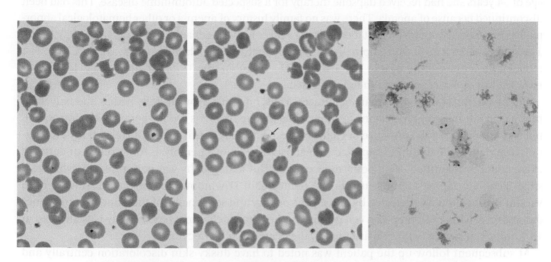

Peak Name	Calibrated Area %	Area %	Retention Time (min)	Peak Area
F	0.2	– – –	1.09	3125
Unknown	– – –	0.9	1.23	13733
P2	– – –	1.1	1.37	15907
P3	– – –	2.4	1.75	35571
A0	– – –	91.2	2.48	1373838
A2	2.7	– – –	3.64	46046
D-window	– – –	1.3	4.06	18895

Total Area: 1,507,117

F Concentration = 0.2 %
A2 Concentration = 2.7 %

Analysis comments:

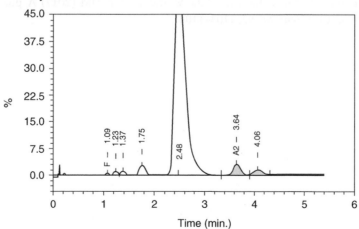

Hematology: 101 Morphology Updates, First Edition. Barbara J. Bain.
© 2023 John Wiley & Sons Ltd. Published 2023 by John Wiley & Sons Ltd.

A 36-year-old Northern European female was referred for a tertiary opinion regarding an unexplained anemia. She had a 6-month history of symptomatic anemia, jaundiced sclerae and dark urine. At the age of 34 years she had received dapsone therapy for a suspected autoimmune disease. This had been discontinued because of anemia. There was no family history of anemia or other hematological abnormality. Her blood count on presentation showed a macrocytic anemia with Hb 88 g/l and MCV 101 fl. The WBC was 6.6×10^9/l with a normal differential count, and the platelet count was 314×10^9/l. The reticulocyte count was elevated to 326×10^9/l (12%). Bilirubin was 33 µmol/l and lactate dehydrogenase 372 iu/l. A direct antiglobulin test was negative.

Her blood film showed irregularly contracted cells, bite cells, Howell–Jolly bodies and polychromasia (top left image). Protrusions from red cells and dense areas within hemighosts (top center, red arrow) suggested the presence of Heinz bodies, this being confirmed on a Heinz body preparation (top right). The findings supported a diagnosis of oxidative hemolysis but no clear underlying cause could be identified. High performance liquid chromatography was performed and showed a small peak eluting at 4.09 minutes in the hemoglobin D window (lower image). However, no variant hemoglobin was detected by mass spectrometry and sequencing of the α and β globin genes did not reveal any mutation. Red cell glucose-6-phosphate dehydrogenase activity was elevated at 13.9 units/g of hemoglobin, consistent with the degree of reticulocytosis.

At subsequent follow-up the patient was noted to have dusky skin discoloration centrally and peripherally. Blood gas analysis revealed a methemoglobin level of 28%. The patient was admitted for further management. Methylene blue treatment was followed by a significant fall of methemoglobin to under 5%, but the reduction was transient and levels rose to over 20% within 24 hours. Red cell cytochrome b5 reductase activity was normal, excluding congenital methemoglobinemia. Acquired methemoglobinemia due to ongoing exposure to an oxidant drug or chemical was suspected. A toxicology screen revealed a significant quantity of dapsone, a potent oxidant, in the patient's urine and blood samples.

The patient was unable to explain the findings but after discussion of these results, the methemoglobin level fell rapidly and Hb together with markers of hemolysis progressively improved. Psychological support was arranged.

Original publication: Morris A, Bain BJ, Atta M and Layton DM (2017) A puzzling case of methemoglobinemia. *Am J Hematol*, **92**, 1103–1104.

95 Hodgkin lymphoma in a bone marrow aspirate

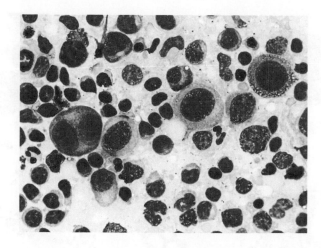

A previously well 60-year-old man presented with a 6-week history of weight loss, drenching night sweats and fevers. His blood count showed Hb 100 g/l, WBC 15 × 10^9/l, neutrophils 10 × 10^9/l, eosinophils 2 × 10^9/l, lymphocytes 0.6 × 10^9/l and platelets 550 × 10^9/l. Computed tomography of the neck, thorax, abdomen and pelvis showed splenomegaly without any lymphadenopathy. A bone marrow aspirate was hypercellular with prominent eosinophils, increased myeloid precursors and increased small lymphocytes and plasma cells. In addition, a population of large cells with plentiful agranular cytoplasm (mononuclear Hodgkin cells) and some binucleate cells (Reed–Sternberg cells) were identified (image). These appearances are characteristic of classic Hodgkin lymphoma. The diagnosis was confirmed by trephine biopsy with the Hodgkin and Reed–Sternberg cells showing the typical immunophenotype of CD15+, CD30+, CD45– and CD20–.

In the era of positron emission tomography staging, bone marrow aspiration and trephine biopsies are now infrequently performed in Hodgkin lymphoma. Even so, it is unusual for Reed–Sternberg and Hodgkin cells to be visualized in an aspirate as the lymphomatous infiltrate generates marrow fibrosis which impairs aspiration. It is important that hematologists retain the ability to recognize the morphology of Hodgkin lymphoma in a bone marrow aspirate on the rare occasions it is encountered.

Original publication: Parsons K, Leach M and Bain BJ (2020) Hodgkin lymphoma in bone marrow aspirate. *Am J Hematol*, **95**, 328–329.

Hematology: 101 Morphology Updates, First Edition. Barbara J. Bain.
© 2023 John Wiley & Sons Ltd. Published 2023 by John Wiley & Sons Ltd.

96 Giant proerythroblasts in pure red cell aplasia due to parvovirus B19 infection in a patient with rheumatoid arthritis

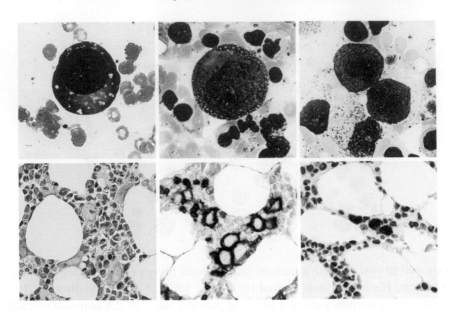

A 72-year-old woman with rheumatoid arthritis presented with fatigue and chest pain and was noted to have an Hb of 43 g/l with absolute reticulocytopenia (3×10^9/l, normal range 50–100). The WBC and platelet counts were normal. The blood film showed no specific features. A bone marrow aspirate showed an absence of normal erythroid precursors with normal granulopoiesis and mega-karyopoiesis. There were a number of extremely large cells, up to 50 μm in diameter with deeply basophilic cytoplasm, occasionally with vacuolation (top left and center images). Some nuclei had inclusions and multiple bare nuclei from these giant cells were also apparent (top right). The bone marrow trephine biopsy confirmed the presence of giant proerythroblasts with an absence of more mature erythroid precursors (bottom left, H&E; bottom center, immunoperoxidase for glycophorin A). Immunoperoxidase for parvovirus viral capsid antigens VP1 and VP2 showed positivity in the nuclei of the giant proerythroblasts (bottom right). No parvovirus B19 IgM or IgG antibodies were detected but PCR for parvovirus DNA was positive.

Parvovirus B19 is a single-stranded DNA virus that specifically infects red cell precursors. Pure red cell aplasia caused by this virus is associated with the presence of giant proerythroblasts, often with viral inclusions ('lantern cells') in the bone marrow. This diagnosis should be considered in immunosuppressed individuals with apparent red cell aplasia as protracted infection and severe anemia can result from their inability to clear the virus. Serology may be negative and PCR analysis is often required for detection of the virus in these patients. The patient reported here was receiving methotrexate, sulphasalazine, hydroxychloroquine and golimumab (antitumor necrosis factor α) therapy for her rheumatoid arthritis. Treatment with intravenous immunoglobulin resulted in reticulocytosis within 2 weeks of administration.

Original publication: Harper K, McIlwaine L, Leach M and Bain BJ (2020) Giant proerythroblasts in pure red cell aplasia due to parvovirus B19 infection in a patient with rheumatoid arthritis. *Am J Hematol*, **95**, 990–991.

Hematology: 101 Morphology Updates, First Edition. Barbara J. Bain.
© 2023 John Wiley & Sons Ltd. Published 2023 by John Wiley & Sons Ltd.

97 A myeloid neoplasm with *FIP1L1::PDGFRA* presenting as acute myeloid leukemia

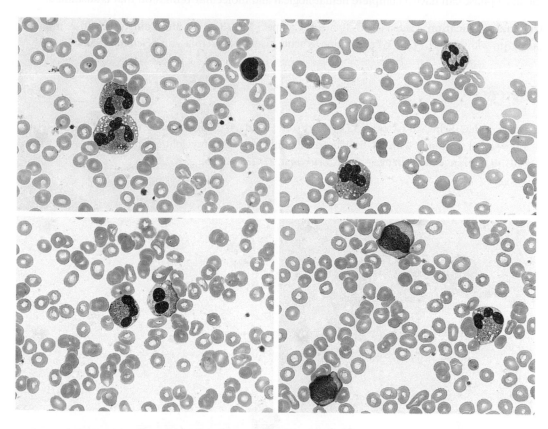

A 30-year-old Spanish woman presented with recurrent respiratory tract infections over several months, together with asthenia and swelling of the abdomen and legs. On examination, she had hepatosplenomegaly, ascites and edema of both legs. Computed tomography confirmed these observations and, in addition, showed bilateral pleural effusions. Her blood count showed WBC 17.45 × 10^9/l, Hb 96 g/l, MCV 93 fl and platelet count 99 × 10^9/l. There was marked eosinophilia (5.2 × 10^9/l) and mild monocytosis (1.05 × 10^9/l) with small numbers of granulocyte precursors and 6% blast cells. Eosinophils showed cytological abnormalities including vacuolation, hypogranularity, hypolobation, hyperlobation and basophilic granules in mature eosinophils (images). Blast cells were medium sized with a high nucleocytoplasmic ratio and fine azurophilic granules (bottom right). A bone marrow aspirate showed 25–30% blast cells and more than 20% eosinophils, these showing similar cytological abnormalities to those in the peripheral blood. Cytogenetic analysis was normal but fluorescence *in situ* hybridization showed the presence of *FIP1L1::PDGFRA*. A diagnosis of myeloid neoplasm with *PDGFRA* rearrangement presenting as acute myeloid leukemia (AML) was made. Subsequent next generation sequencing showed two subclonal mutations in *BCORL1* (variant allele frequency 8.4%) and *PTPN11* (variant allele frequency 7.4%).

The patient was treated with imatinib in an initial dose of 400 mg daily for 2 weeks and then 200 mg daily, without chemotherapy, and entered rapid remission. After 4 months on 200 mg daily of imatinib she was well with a normal blood count.

Hematology: 101 Morphology Updates, First Edition. Barbara J. Bain.
© 2023 John Wiley & Sons Ltd. Published 2023 by John Wiley & Sons Ltd.

Because of their sensitivity to tyrosine kinase inhibitors, the identification of neoplasms with a rearrangement of *PDGFRA* or *PDGFRB* is of considerable importance. Even patients presenting in the blast phase can have a complete hematological and molecular remission that is sustained.[1]

Original publication: Cadenas FL, Bua BR, Campelo MD, Rieu JB and Bain BJ (2020) A myeloid neoplasm with *FIP1L1-PDGFRA* presenting as acute myeloid leukemia. *Am J Hematol*, **95**, 1214–1215.

Reference

1 Metzgeroth G, Schwaab J, Gosenca D, Fabarius A, Haferlach C, Hochhauz A *et al.* (2013) Long-term follow-up of treatment with imatinib in eosinophilia-associated myeloid/lymphoid neoplasms with PDGFR rearrangements in blast phase. *Leukemia*, **27**, 2254–2256.

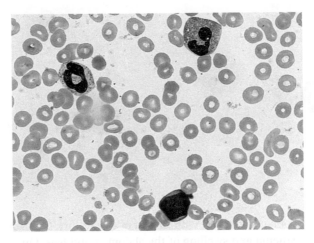

A further image showing a blast cell and an eosinophil with occasional vacuoles.

98   Breast implant-associated anaplastic large cell lymphoma

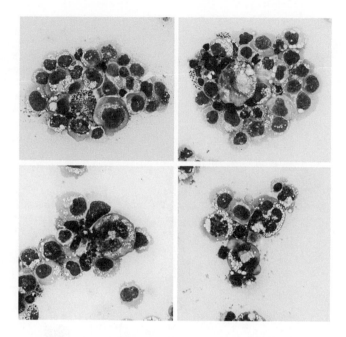

A 39-year-old woman with bilateral textured silicone breast implants *in situ* for 5 years, presented with a short history of right breast swelling with erythema and axillary pain. Magnetic resonance imaging demonstrated a large seroma around the implant but there was no breast or soft tissue mass lesion and no lymphadenopathy. The fluid was subsequently sampled and cytospin preparations showed a population of pleomorphic lymphoid cells with prominent vacuolation of the basophilic cytoplasm (images). Many cells were undergoing karyorrhexis (notable in the top left and top right images). Notably, there are a number of larger cells with 'horse-shoe' or reniform-shaped nuclei (all images); these are 'hallmark cells', which are typically identified in anaplastic large cell lymphoma. Flow cytometric immunophenotyping showed these cells to express CD4, CD5, CD7 and CD30. This was confirmed on immunohistochemistry on a cell pellet but ALK (anaplastic lymphoma kinase) was not expressed.

Breast implant-associated anaplastic large cell lymphoma (BIA-ALCL) is a rare but well-recognized condition, noted as a provisional entity in the 2016 *WHO Classification of Tumours of Haematopoietic and Lymphoid Tissues* and as a definitive entity in the 2022 classification. The condition appears to arise as a result of chronic inflammation around textured breast implants and typically the first manifestation is the development of a seroma. Although spread can occur elsewhere, BIA-ALCL is almost always localized to the implant capsule. Importantly, when the implant is removed this frequently leads to resolution of the lymphoma. Patients with more advanced stage disease have a much less favorable prognosis and may need systemic chemotherapy. Early recognition of this condition and knowledge of its association with breast prostheses is essential in hematological practice.

Original publication: Hopkins D, Smyth D, Leach M and Bain BJ (2022) Breast implant-associated anaplastic large cell lymphoma. *Am J Hematol*, **97**, 1257–1258.

99 Large granular lymphocytosis induced by dasatinib

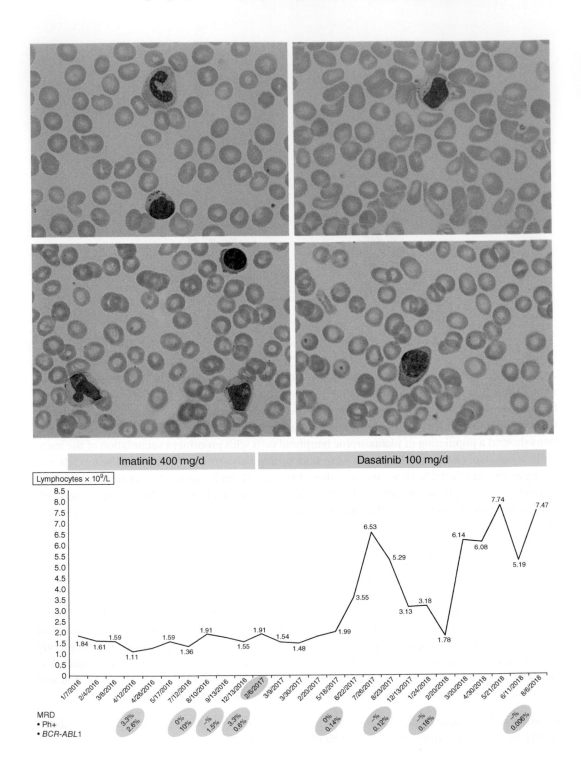

Hematology: 101 Morphology Updates, First Edition. Barbara J. Bain.
© 2023 John Wiley & Sons Ltd. Published 2023 by John Wiley & Sons Ltd.

A diagnosis of chronic myeloid leukemia, *BCR::ABL1* positive, was made in a 66-year-old woman in 2016. She was treated with imatinib, 400 mg daily, with initial response but 12 months later the cytogenetic response had been lost. Therapy was therefore changed to dasatinib, 100 mg/day. Six months after the change of therapy her lymphocyte count, which had been normal, had increased to 6.54×10^9/l and by 18 months had risen further with peak counts of 7.5–7.7×10^9/l (graph).

The patient's blood film showed that the lymphocytosis was largely due to an increase of large granular lymphocytes. These were mainly cytologically normal, often with large, prominent granules (top images), but a minority showed dysplastic features, such as nuclear lobulation (bottom left image). A minor population showed features resembling those seen in viral infections (bottom right image). A bone marrow aspirate showed 28% lymphocytes, mainly small lymphocytes, which had less obvious cytoplasmic granules. Peripheral blood immunophenotyping showed 24% of leucocytes to have the immunophenotype of natural killer (NK) cells: CD3–/CD16+/CD56+; most expressed CD2 but 9% of this population did not. Molecular evaluation showed an IS (International Scale) BCR-ABL1 of 0.06%, the lowest value obtained since the beginning of dasatinib therapy.

Large granular lymphocytosis is a well-recognized feature of dasatinib therapy. It may occur in as many as a third to a half of patients on this drug.[1,2] Both T cells and NK cells can be increased. Rearrangement of T-cell receptor genes can indicate monoclonality or oligoclonality of both T cells and NK cells with a small clone being present at diagnosis and expanding during dasatinib therapy.[1] This phenomenon may be associated with higher response rates, longer response durations and increased overall survival.[1,2]

Original publication: Fernandes F, Ramalho R, Barreira R, Silveira M and Bain BJ (2021) Large granular lymphocytosis induced by dasatinib. *Am J Hematol*, **96**, 395–396.

References

1 Kreutzman A, Juvonen V, Kairisto V, Ekblom M, Stenke L, Seggewiss R *et al.* (2010) Mono/oligoclonal T and NK cells are common in chronic myeloid leukemia patients at diagnosis and expand during dasatinib therapy. *Blood*, **116**, 772–782.

2 Schiffer CA, Cortes JE, Hochhaus A, Saglio G, le Coutre P, Porkka K *et al.* (2016) Lymphocytosis after treatment with dasatinib in chronic myeloid leukemia: effects on response and toxicity. *Cancer*, **122**, 1398–1407.

100 The distinctive cytology and disease evolution of blastic plasmacytoid dendritic cell neoplasm

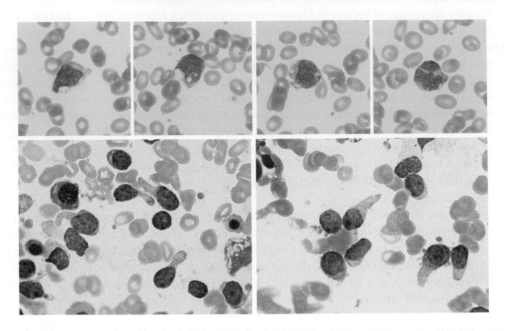

A 26-year-old Portuguese man presented with a 2-week history of oral ulcers (tongue and lower lip) and was found to have profound neutropenia. He had a history of celiac disease. His brother had died abroad 2 years earlier at the age of 30 years with a diagnosis of 'leukemia of mixed phenotype' (CD34+, CD117+, CD33 weak, CD7 weak, CD19+, HLA-DR weak and CD45 weak).

The patient's blood count showed WBC 4.26×10^9/l, Hb 106 g/l, platelets 156×10^9/l, neutrophils 0.34×10^9/l, monocytes 1.53×10^9/l and abnormal mononuclear cells 0.17×10^9/l. The abnormal mononuclear cells (top images above) had basophilic cytoplasm, which was often vacuolated and sometimes had an irregular margin; chromatin showed some condensation and a proportion of cells had indistinct nucleoli. A bone marrow aspirate was hypercellular with 74% abnormal cells. These showed a spectrum of cytological features. There were prominent medium-sized cells with round nuclei showing some chromatin condensation, and agranular, basophilic cytoplasm; a distinctive feature was the presence of broad cytoplasmic tails in a significant proportion of cells (lower images above). In addition, there were small cells with a high nucleocytoplasmic ratio and basophilic cytoplasm. A minority of cells of both types had fine cytoplasmic vacuoles. The small cells resembled those in the peripheral blood but vacuolation was less prominent. Cytochemistry was negative for myeloperoxidase and α naphthyl acetate esterase (ANAE). Periodic acid–Schiff (PAS) staining showed coarse granular PAS positivity in some cells. Flow cytometric immunophenotyping showed two pathological populations of cells. The less mature population (30% of cells) expressed CD34, CD117, CD13, weak CD33, HLA-DR, CD7, CD123 and weak CD45; they were negative for CD4, CD56, myeloperoxidase and other myeloid and B- and T-lineage markers. The second population (50% of cells) expressed CD36, HLA-DR, CD7, CD4, weak CD45 and strong CD123; they were negative for CD56, myeloperoxidase and other myeloid and B- and T-lineage markers. Fluorescence *is situ* hybridization showed no rearrangement of *KMT2A*, *CBFB* or *RUNX1* and cytogenetic analysis was normal. A diagnosis of blastic plasmacytoid dendritic cell neoplasm (BPDCN) was made.

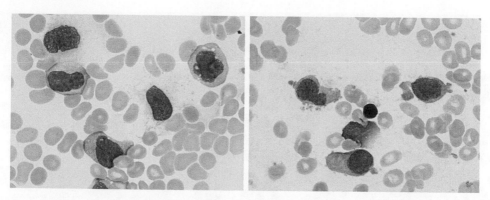

The patient received intensive chemotherapy (hyper-CVAD regime) and was further evaluated after 28 days. The bone marrow aspirate now showed replacement by pleomorphic neoplastic cells with basophilic, sometimes vacuolated, cytoplasm, some cells with features suggesting monocytic differentiation (images above) but with there being some residual cells with blunt cytoplasmic tails (right). Immunophenotyping showed less than 1% CD34+ cells. Cells consistent with immature blastoid dendritic cells (CD117 variable, CD4 weak, CD123++, HLA-DR+ and CD56−) constituted 8% of cells, while 44% were maturing cells of monocyte lineage (CD14+, CD36+, HLA-DR+ and CD56 weak). There were now 65% ANAE+ cells.

The characteristic immunophenotype of BPDCN is positivity for CD4 and CD56. In the small minority of patients in whom one of these markers is negative, the diagnosis is strengthened by the expression of CD123, TCL1A or CD303. As in our patient, the distinctive cytology, particularly the cytoplasmic tails or pseudopodia-like extensions, also strongly suggests this diagnosis.[1,2] In one series, 15 or 23 patients showed this feature.[1] Vacuoles are also often present[1] and have been attributed to the presence of glycogen. We have previously observed, as in this patient, that the most diagnostic cells are those in the bone marrow.

There is an association between BPDCN and myeloid neoplasms, particularly chronic myelomonocytic leukemia. We postulate that in this patient there was a common stem cell origin with chemotherapy being associated with a regression of the most primitive blastic plasmacytoid dendritic cell population with a shift of differentiation towards the monocyte lineage, shown by changes in the morphology, cytochemistry and immunophenotype.

Original publication: Fernandes F, Barreira R, Cortez J, Silveira M and Bain BJ (2018) The distinctive cytology and disease evolution of blastic plasmacytoid dendritic cell neoplasm. *Am J Hematol*, **93**, 1431–1432.

References

1 Feuillard J, Jacob MC, Valensi F, Maynadié M, Gressin R, Chaperot L *et al.* (2002) Clinical and biologic features of CD4(+)CD56(+) malignancies. *Blood*, **99**, 1556–1563.
2 Kassam S, Rice A, Morilla R and Bain BJ (2007) Teaching case, Case 35: an unusual haematological neoplasm characterized by cells with cytoplasmic tails. *Leuk Lymphoma*, **48**, 1208–1210.

101 Platelet phagocytosis as a cause of pseudothrombocytopenia

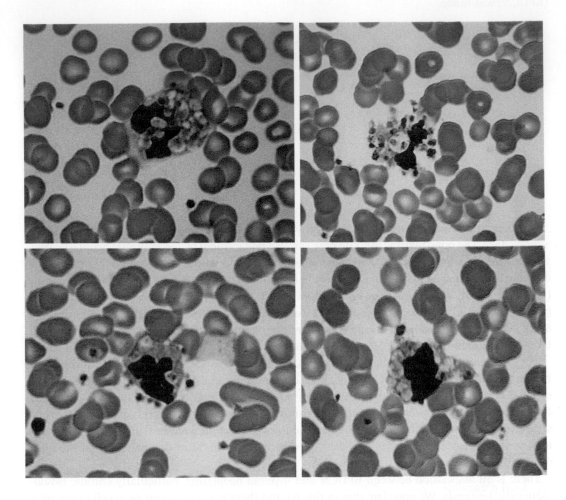

A 43-year-old man with no history of bleeding or bruising had a routine blood count performed prior to angiography. The automated platelet count on an impedance-based instrument was 108 × 10^9/l. The rest of the blood count was normal.

Because of the unexpected thrombocytopenia, a blood film was examined. This showed extensive phagocytosis of the platelets (images). There was limited platelet satellitism, suggesting that this may have been the first stage of the process (bottom left). In some cells, phagocytic vacuoles were clearly apparent (top right). An *in vitro* phenomenon was suspected. A repeat blood sample was therefore taken and a film prepared with no exposure to anticoagulant showed no phagocytosis. After a brief exposure to EDTA occasional neutrophils showed satellitism and phagocytosis and the platelet count was 158 × 10^9/l. By 4 hours, phagocytosis was extensive. The patient proceeded to an uneventful angiogram.

Original publication: Campbell V, Fosbury E and Bain BJ (2009) Platelet phagocytosis as a cause of pseudothrombocytopenia. *Am J Hematol*, **84**, 362.

Hematology: 101 Morphology Updates, First Edition. Barbara J. Bain.
© 2023 John Wiley & Sons Ltd. Published 2023 by John Wiley & Sons Ltd.

Test yourself

Multiple choice questions: one to five answers may be correct.

Question 1

In paroxysmal cold hemoglobinuria, a direct antiglobulin test is likely to show:

A Complement
B IgG
C IgG plus IgM
D IgM
E No immunoglobulin or complement

Question 2

Basophilic stippling can be a prominent feature of:

A Arsenic poisoning
B Beta thalassemia heterozygosity
C Lead poisoning
D Hereditary elliptocytosis
E Pyrimidine 5′ nucleotidase deficiency

Question 3

Prominent Howell–Jolly bodies are likely in:

A Celiac disease with folic acid deficiency
B Hypersplenism
C Megaloblastic anemia post-splenectomy
D Methotrexate therapy
E Pernicious anemia

Hematology: 101 Morphology Updates, First Edition. Barbara J. Bain.
© 2023 John Wiley & Sons Ltd. Published 2023 by John Wiley & Sons Ltd.

Question 4

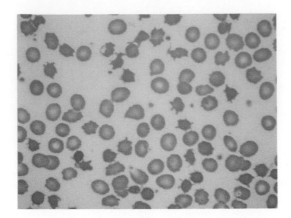

The morphological abnormality in this patient with neurological abnormalities is:

A Acanthocytosis
B Bite cells
C Crenation
D Echinocytosis
E Pyknocytosis

Question 5

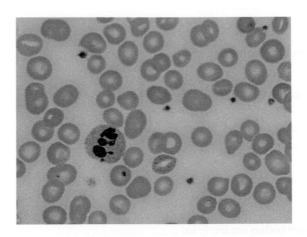

The blood film illustrated shows:

A A Howell–Jolly body
B A nuclear drumstick
C Basophilic stippling
D Pappenheimer bodies
E Thrombocytopenia

Question 6

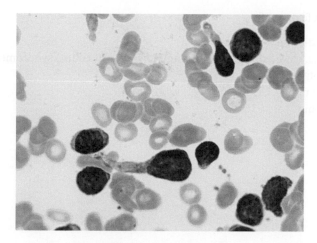

This bone marrow film is from a patient with mild anemia. The more mature of the abnormal cells expressed CD45, CD4, CD7 and CD123 (strong) but not CD56 or myeloperoxidase. The most likely diagnosis is:

A Acute myeloid leukemia
B Blastic plasmacytoid dendritic cell neoplasm
C Mixed phenotype acute leukemia
D Non-hematopoietic neoplasm
E Systemic mastocytosis

Question 7

Irregularly contracted cells are expected in:

A Autoimmune hemolytic anemia
B Hemoglobin C disease
C Hemolysis in glucose-6-phosphate dehydrogenase (G6PD) deficiency
D The majority of patients with β thalassemia heterozygosity
E The presence of an unstable hemoglobin

Question 8

Blood film features indicating the likelihood of pseudothrombocytopenia, rather than true thrombocytopenia, include:

A Cryoglobulin deposits
B Fibrin strands
C Numerous giant platelets
D Phagocytosis of platelets
E Platelet satellitism

Question 9

Schistocytes can be a feature of:

A Graft-versus-host disease
B HELLP (hemolysis, elevated liver enzymes and low platelet count) syndrome
C Mechanical hemolytic anemia
D Megaloblastic anemia
E Sideroblastic anemia

Question 10

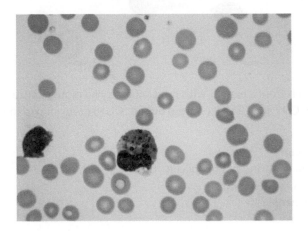

This blood film is from a 58-year-old woman with acute myelomonocytic leukemia. The most likely cytogenetic/genetic abnormality is:

A inv(16)(p13.1q22) or t(16;16)(p13.1;q22)/*CBFB::MYH11*
B *NPM1* mutated + *FLT3* internal tandem duplication
C t(1;22)(p13.3;q13.1)/*RBM1::MKL1*
D t(8;16)(p11;p13)/*KAT6A::CREBBP*
E t(15;17)(q24.1;q21.2)/*PML::RARA*

Question 11

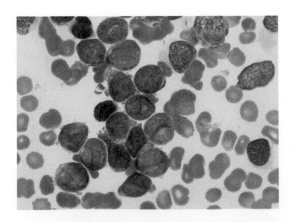

This bone marrow film is from a 69-year-old woman with fever with a WBC of 151.5 × 10⁹/l and a platelet count of 33 × 10⁹/l. The peripheral blood film showed numerous blast cells. The most likely cytogenetic/genetic abnormality is:

A inv(16)(p13.1q22) or t(16;16)(p13.1;q22)/*CBFB::MYH11*
B *NPM1* mutated + *FLT3* internal tandem duplication
C t(1;22)(p13.3;q13.1)/*RBM1::MKL1*
D t(8;16)(p11;p13)/*KAT6A::CREBBP*
E t(15;17)(q24.1;q21.2)/*PML::RARA*

Question 12

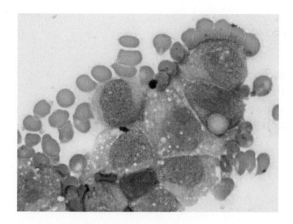

This bone marrow film is from a patient who presented with hemorrhagic manifestations as a result of disseminated intravascular coagulation. Immunophenotyping of the blast cells indicated early monocytic differentiation. There was expression of HLA-DR but not CD34. The most likely cytogenetic/genetic abnormality is:

A inv(16)(p13.1q22) or t(16;16)(p13.1;q22)/*CBFB::MYH11*
B *NPM1* mutated + *FLT3* internal tandem duplication
C t(1;22)(p13.3;q13.1)/*RBM1::MKL1*
D t(8;16)(p11;p13)/*KAT6A::CREBBP*
E t(15;17)(q24.1;q21.2)/*PML::RARA*

Question 13

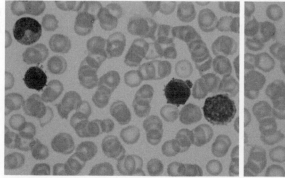
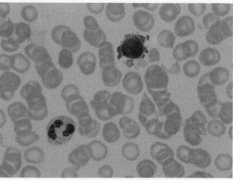

This blood film is from a phenotypically normal neonate presenting with hepatosplenomegaly and bleeding, who was found to have an increased WBC, circulating blast cells and thrombocytopenia. The most likely cytogenetic/genetic abnormality is:

A inv(16)(p13.1q22) or t(16;16)(p13.1;q22)/*CBFB::MYH11*
B *NPM1* mutated + *FLT3* internal tandem duplication
C t(1;22)(p13.3;q13.1)/*RBM1::MKL1*
D t(8;16)(p11;p13)/*KAT6A::CREBBP*
E t(15;17)(q24.1;q21.2)/*PML::RARA*

Question 14

Microspherocytes are expected in:

A Burns
B Delayed transfusion reaction
C Hemolytic disease of the newborn
D Mechanical hemolytic anemia
E Microangiopathic hemolytic anemia

Question 15

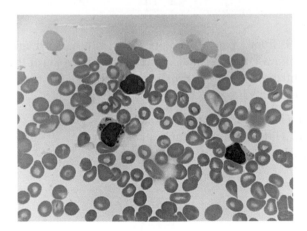

This blood film from a young boy with leucocytosis, anemia and thrombocytopenia was associated with an unexpected immunophenotype. There was expression of CD45 (weak), CD34, CD10, CD19, CD20, cytoplasmic (c) CD79a, CD33 and HLA-DR. Myeloperoxidase was not expressed. The diagnosis is:

A Acute lymphoblastic leukemia
B Acute myeloid leukemia
C Mixed phenotype acute leukemia
D NK cell leukemia
E T-cell large granular lymphocytic leukemia

Question 16

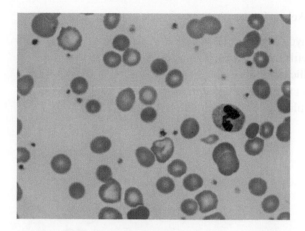

This is the blood film of a middle-aged Northern European woman who has recently presented with anemia. Her Hb is 68 g/l, MCV 110 fl and MCHC 306 g/l. The most likely diagnosis is:

A Acute hemolysis in G6PD deficiency
B Cold hemagglutinin disease
C Hereditary spherocytosis
D Megaloblastic anemia
E Warm autoimmune hemolytic anemia

Question 17

The blood film may show features of hyposplenism in:

A Amyloidosis
B Celiac disease
C Portal hypertension
D Sickle cell anemia
E Splenic infiltration

Question 18

The blood film may reveal that an automated platelet count is falsely high as the result of the presence of:

A Döhle bodies
B Fragments of white cells
C Gray platelets
D Macrothrombocytes
E Schistocytes

Question 19

Toxic granulation is likely to be seen in:

A Atypical chronic myeloid leukemia, *BCR::ABL1* negative
B Chronic myeloid leukemia, *BCR::ABL1* positive
C Chronic neutrophilic leukemia
D Leukemoid reaction to multiple myeloma
E Therapy with granulocyte colony-stimulating factor (G-CSF)

Question 20

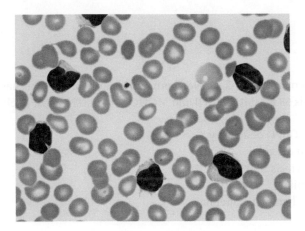

The image shows the blood film of a middle-aged man with lymphadenopathy who had been found to have lymphocytosis and mild anemia. Immunophenotyping showed clonal B cells. The specific marker that would support your diagnostic suspicion is:

A CD5
B CD10
C CD23
D CD34
E Cyclin D1

Answers to test cases

Question 1

In paroxysmal cold hemoglobinuria, a direct antiglobulin test is likely to show:

A Complement [True]
B IgG
C IgG plus IgM
D IgM
E No immunoglobulin or complement

The correct answer is A. Although the responsible antibody is an IgG antibody, the direct anti-globulin test detects only complement. The antibody fixes complement on cooling but on rewarming detaches from the red cell membrane so that only complement is detected. See Update 8.

Question 2

Basophilic stippling can be a prominent feature of:

A Arsenic poisoning [True]
B Beta thalassemia heterozygosity [True]
C Lead poisoning [True]
D Hereditary elliptocytosis
E Pyrimidine 5′ nucleotidase deficiency [True]

The correct answers are A, B, C and E. Basophilic stippling can be an important clue to the diagnosis of lead poisoning (as in this case) and pyrimidine 5′ nucleotidase deficiency. However, it is a non-specific feature with β thalassemia heterozygosity and arsenic poisoning, being among the many other conditions in which it can be found. It is seen in only a minority of cases of β thalassemia heterozygosity. See Updates 10 and 15.

Question 3

Prominent Howell–Jolly bodies are likely in:

A Celiac disease with folic acid deficiency [True]
B Hypersplenism
C Megaloblastic anemia post-splenectomy [True]
D Methotrexate therapy
E Pernicious anaemia

The correct answers are A and C. Prominent Howell–Jolly bodies are seen when there is a combination of megaloblastic anemia, leading to dyserythropoiesis and increased formation of Howell–Jolly bodies, and splenic absence of atrophy, leading to a reduced rate of clearance. Celiac disease leads to both splenic atrophy and megaloblastic anemia due to folic acid deficiency or, less often, vitamin B_{12} deficiency. See Update 5.

Question 4

The morphological abnormality in this patient with neurological abnormalities is:

A Acanthocytosis [True]
B Bite cells
C Crenation
D Echinocytosis
E Pyknocytosis

The correct answer is A. This is another example of the acanthocytosis seen in a patient with neuro-acanthocytosis. See Update 9.

Question 5

The blood film illustrated shows:

A A Howell–Jolly body
B A nuclear drumstick [True]
C Basophilic stippling [True]
D Pappenheimer bodies
E Thrombocytopenia

The correct answers are B and C. There is a platelet overlying a red cell but no Howell–Jolly bodies. The nuclear drumstick indicates that this is the blood film of a female. The red cell inclusions have the characteristics of basophilic stippling not Pappenheimer bodies. This is another image of the film of the patient with lead poisoning in Update 10.

Question 6

This bone marrow film is from a patient with mild anemia. The more mature of the abnormal cells expressed CD45, CD4, CD7 and CD123 (strong) but not CD56 or myeloperoxidase. The most likely diagnosis is:

A Acute myeloid leukemia
B Blastic plasmacytoid dendritic cell neoplasm [True]

C Mixed phenotype acute leukemia
D Non-hematopoietic neoplasm
E Systemic mastocytosis

The correct answer is B. Consideration of both the cytological features and the immunophenotype leads to a diagnosis of blastic plasmacytoid dendritic cell neoplasm. Systemic mastocytosis can also have cells with cytoplasmic tails but the neoplastic cells tend to be more spindle-shaped (see Update 88) and furthermore the cells illustrated in this case appear to be agranular. The strong CD123 positivity points to the correct diagnosis. CD56 is usually, but not always, positive. See Update 100.

Question 7

Irregularly contracted cells are expected in:

A Autoimmune hemolytic anemia
B Hemoglobin C disease [True]
C Hemolysis in glucose-6-phosphate dehydrogenase (G6PD) deficiency [True]
D The majority of patients with β thalassemia heterozygosity
E The presence of an unstable hemoglobin [True]

The correct answers are B, C and E. Irregularly contracted cells are characteristic of oxidant damage, particularly but not only when there is hemolysis in G6PD deficiency, and also in hemoglobin C disease, hemoglobin E disease and in the presence of an unstable hemoglobin. Only a minority of patients with β thalassemia heterozygosity show this feature. See Updates 6, 7, 60, 69, 91, 92 and 94.

Question 8

Blood film features indicating the likelihood of pseudothrombocytopenia, rather than true thrombocytopenia, include:

A Cryoglobulin deposits
B Fibrin strands [True]
C Numerous giant platelets [True]
D Phagocytosis of platelets [True]
E Platelet satellitism [True]

The correct answers are B, C, D and E. Fibrin strands indicate that clotting has occurred in the sample and platelets are likely to have been consumed. Platelet satellitism and phagocytosis lead to the platelets passing though the orifice of an automated counter together with neutrophils so that they are not identifiable by cell sizing technology. Giant platelets are under-counted as they fall above the threshold of the automated instrument.

Question 9

Schistocytes can be a feature of:

A Transplant-associated thrombotic microangiopathy [True]
B HELLP (hemolysis, elevated liver enzymes and low platelet count) syndrome [True]
C Mechanical hemolytic anemia [True]
D Megaloblastic anemia [True]
E Sideroblastic anemia

The correct answers are A, B, C and D. Schistocytes can result from microangiopathic and mechanical hemolytic anemias. They can also result from dyserythropoiesis, as in severe megaloblastic anemia, and this has led to confusion with thrombotic thrombocytopenic purpura. They are not a feature of sideroblastic anemia. See Updates 4, 12, 48 and 49.

Question 10

This blood film is from a 58-year-old woman with acute myelomonocytic leukemia. The most likely cytogenetic/genetic abnormality is:

A inv(16)(p13.1q22) or t(16;16)(p13.1;q22)/*CBFB::MYH11* [True]
B *NPM1* mutated + *FLT3* internal tandem duplication
C t(1;22)(p13.3;q13.1)/*RBM1::MKL1*
D t(8;16)(p11;p13)/*KAT6A::CREBBP*
E t(15;17)(q24.1;q21.2)/*PML::RARA*

The correct answer is A. The film in this patient with acute myelomonocytic leukemia shows an eosinophil precursor with large proeosinophilic granules, which have basophilic staining characteristics. This suggests the correct diagnosis. See Update 44.

Question 11

This bone marrow film is from a 69-year-old woman with fever, a WBC of 151.5×10^9/l and a platelet count of 33×10^9/l. The peripheral blood film showed numerous blast cells. The most likely cytogenetic/genetic abnormality is:

A inv(16)(p13.1q22) or t(16;16)(p13.1;q22)/*CBFB::MYH11*
B *NPM1* mutated + *FLT3* internal tandem duplication [True]
C t(1;22)(p13.3;q13.1)/*RBM1::MKL1*
D t(8;16)(p11;p13)/*KAT6A::CREBBP*
E t(15;17)(q24.1;q21.2)/*PML::RARA*

The correct answer is B. There are at least three blast cells showing a large indentation in the nucleus, creating blast cells with cup-shaped nuclei. Although not specific, this feature is most characteristic of AML with both an *NPM1* mutation and *FLT3*-ITD. See Update 59.

Question 12

This bone marrow film is from a patient who presented with hemorrhagic manifestations as a result of disseminated intravascular coagulation. Immunophenotyping of the blast cells indicated early monocytic differentiation. There was expression of HLA-DR but not CD34. The most likely cytogenetic/genetic abnormality is:

A inv(16)(p13.1q22) or t(16;16)(p13.1;q22)/*CBFB::MYH11*
B *NPM1* mutated + *FLT3* internal tandem duplication
C t(1;22)(p13.3;q13.1)/*RBM1::MKL1*

D t(8;16)(p11;p13)/*KAT6A::CREBBP* [True]
E t(15;17)(q24.1;q21.2)/*PML::RARA*

The correct answer is D. The distinctive feature of the monoblasts/promonocytes that dominate the marrow in this patient is an erythrophagocytic blast cell. This subtype of AML is characterized by DIC and hemophagocytosis by the leukemic cells. In contrast to acute promyelocytic leukemia, in which usually neither HLA-DR nor CD34 is expressed, this subtype of AML usually shows expression of HLA-DR but not CD34. This case is a reminder that acute promyelocytic leukemia is not the only subtype of AML that often has DIC. See Update 83.

Question 13

This blood film is from a phenotypically normal neonate presenting with hepatosplenomegaly and bleeding, who was found to have an increased WBC, circulating blast cells and thrombocytopenia. The most likely cytogenetic/genetic abnormality is:

A inv(16)(p13.1q22) or t(16;16)(p13.1;q22)/*CBFB::MYH11*
B *NPM1* mutated + *FLT3* internal tandem duplication
C t(1;22)(p13.3;q13.1)/*RBM1::MKL1* [True]
D t(8;16)(p11;p13)/*KAT6A::CREBBP*
E t(15;17)(q24.1;q21.2)/*PML::RARA*

The correct answer is C. The blast cells have the characteristics of megakaryoblasts with budding basophilic cytoplasm. This strongly suggests acute megakaryoblastic leukemia and in a neonate suggests the possibility of t(1;22)(p13.3;q13.1)/*RBM1::MKL1*. A diagnosis of transient abnormal myelopoiesis of Down syndrome might also be considered. See Update 67.

Question 14

Microspherocytes are expected in:

A Burns [True]
B Delayed transfusion reaction
C Hemolytic disease of the newborn
D Mechanical hemolytic anemia [True]
E Microangiopathic hemolytic anemia [True]

The correct answers are A, D and E. Microspherocytes are expected in burns (due to budding off of part of the damaged red cell) and in microangiopathic and mechanical hemolytic anemias (due to fragmentation of red cells). The other conditions are characterized by the presence of spherocytes. See Update 48.

Question 15

This blood film from a young boy with leucocytosis, anemia and thrombocytopenia was associated with an unexpected immunophenotype. There was expression of CD45 (weak), CD34, CD10,

CD19, CD20, cytoplasmic (c) CD79a, CD33 and HLA-DR. Myeloperoxidase was not expressed. The diagnosis is:

A Acute lymphoblastic leukemia [True]
B Acute myeloid leukemia
C Mixed phenotype acute leukemia
D NK cell leukemia
E T-cell large granular lymphocytic leukemia

The correct answer is A. The delicate chromatin pattern indicates that these are blast cells. Despite the prominent granules, the immunotype shows that these are immature B cells expressing CD34 (an indicator of immaturity) and B-cell markers (CD10, CD19, CD20 and cCD79a). The diagnosis is therefore acute lymphoblastic leukemia. CD33 is a myeloid marker but in the absence of myeloperoxidase expression the diagnosis is not mixed phenotype acute leukemia. See Update 31.

Question 16

This is the blood film of a middle-aged Northern European woman who has recently presented with anemia. Her Hb is 68 g/l, MCV 110 fl and MCHC 306 g/l. The most likely diagnosis is:

A Acute hemolysis in G6PD deficiency
B Cold hemagglutinin disease
C Hereditary spherocytosis
D Megaloblastic anemia
E Warm autoimmune hemolytic anemia [True]

The correct answer is E. There are spherocytes rather than irregularly contracted cells; this and the ethnic origin make G6PD deficiency unlikely. There are no agglutinates to suggest cold agglutinin disease. The high MCV is due to marked reticulocytosis not megaloblastic anemia. The apparent recent onset suggests a diagnosis of autoimmune hemolytic anemia rather than hereditary spherocytosis. See Update 71.

Question 17

The blood film may show features of hyposplenism in:

A Amyloidosis [True]
B Celiac disease [True]
C Portal hypertension
D Sickle cell anemia [True]
E Splenic infiltration [True]

The correct answers are A, B, D and E. Portal hypertension does not cause hyposplenism, which can result not only from splenectomy and congenital absence of the spleen but also from (i) replacement of normal splenic tissue by amyloid and when there is infiltration by neoplastic cells; (ii) splenic atrophy (as in celiac disease); (iii) splenic infarction (as in sickle cell disease); and (iv) overload of splenic macrophages in acute hemolytic anemia. See Updates 5, 32 and 71.

Question 18

The blood film may reveal that an automated platelet count is falsely high as the result of the presence of:

A Döhle bodies
B Fragments of white cells [True]
C Gray platelets
D Macrothrombocytes
E Schistocytes [True]

The correct answers are B and E. A falsely high platelet count can be the result of the presence of fragments of red cells (schistocytes) or fragments of the cytoplasm of leukemic cells or lymphoma cells. See Update 23.

Question 19

Toxic granulation is likely to be seen in:

A Atypical chronic myeloid leukemia, *BCR::ABL1* negative
B Chronic myeloid leukemia, *BCR::ABL1* positive
C Chronic neutrophilic leukemia [True]
D Leukemoid reaction to multiple myeloma [True]
E Therapy with granulocyte colony-stimulating factor (G-CSF) [True]

The correct answers are C, D and E. Toxic granulation is best known as an indicator of bacterial infection but there are less common and rare causes. In chronic neutrophilic leukemia it results from a mutation in the gene encoding the receptor for G-CSF, while in a leukemoid reaction to multiple myeloma it results from G-CSF production by the myeloma cells. See Updates 13, 70, 74 and 75.

Question 20

The image shows the blood film of a middle-aged man with lymphadenopathy who had been found to have lymphocytosis and mild anemia. Immunophenotyping showed clonal B cells. The specific marker that would support your diagnostic suspicion is:

A CD5
B CD10 [True]
C CD23
D CD34
E Cyclin D1

The correct answer is B. The film shows mature lymphocytes with deep, narrow nuclear clefts suggesting a diagnosis of follicular lymphoma. Expression of CD10 would support this diagnosis. Expression of CD5 would be expected in chronic lymphocytic leukemia and mantle cell lymphoma. Expression of CD23 would also support the diagnosis of CLL and cyclin D1 is expressed in the majority of patients with mantle cell lymphoma. Since these appear to be mature cells, expression of CD34 would not be expected. See Update 86.

Index

Hematology: 101 Morphology Updates, First Edition. Barbara J. Bain.
© 2023 John Wiley & Sons Ltd. Published 2023 by John Wiley & Sons Ltd.

Printed and bound by CPI Group (UK) Ltd, Croydon, CR0 4YY

18/09/2024

14559428-0001